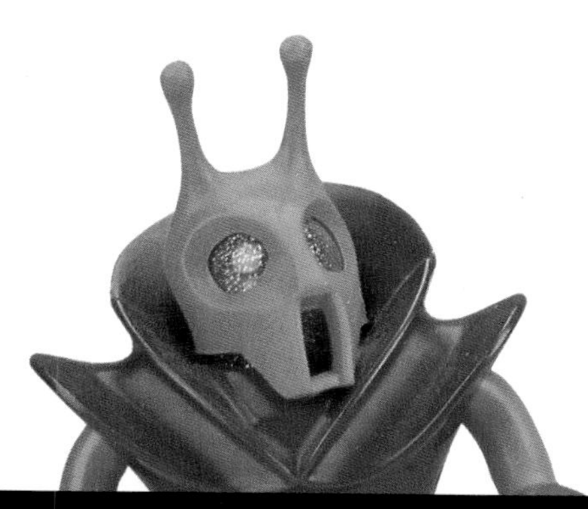

Action Figures of the 1960s

John Marshall

Schiffer Publishing Ltd®

4880 Lower Valley Road, Atglen, PA 19310 USA

Copyright Notice

The Sears catalog pages reproduced in this book are taken from the 1965 and 1966 Christmas catalogs. We are grateful and thrilled that Sears has allowed them to be used in this book.

Library of Congress Catalog Card Number: 98-85390

Book design by Blair Loughrey
Type set in Dom Bold BT/Zurich BT

ISBN: 0-7643-0428-3
Printed in Hong Kong
1 2 3 4

Published by Schiffer Publishing Ltd.
4880 Lower Valley Road
Atglen, PA 19310
Phone: (610) 593-1777; Fax: (610) 593-2002
E-mail: Schifferbk@aol.com
Please write for a free catalog.
This book may be purchased from the publisher.
Please include $3.95 for shipping.

In Europe, Schiffer books are distributed by
Bushwood Books
6 Marksbury Avenue
Kew Gardens
Surrey TW9 4JF England
Phone: 44 (0) 181 392-8585; Fax: 44 (0) 181 392-9876
E-mail: Bushwd@aol.com

Please try your bookstore first.

We are interested in hearing from authors
with book ideas on related subjects.

Contents

Acknowledgments

I did it all by myself.

Just kidding.

As I say in all my books, there would be no book without Scott Talis, owner of the collectibles shop Play With This. You are enjoying the third book to which he has contributed invaluable materials and knowledge. Stop by the shop in the Pennsauken Mart, Pennsauken, New Jersey 08110 (609) 486-4556.

Our second biggest contributor is new to John Marshalldom, but his additions provided many of the rarest items you'll see in these pages. He's Jim Adezio, the renowned toy dealer—and maybe the coolest guy in the business. Visit his Toyrareum, 1101 Asbury Avenue, in Ocean City, New Jersey 08226. (609) 391-0480.

A third dealer/authority, who was especially helpful in pricing out the pieces in the Captain Action chapter, is Tim Welsh. Tim specializes in 1960s and '70s items and can be reached at actiondr@superlink.net.

Fred Mahn, whose name you'll recognize from my other two books, was a key player in providing many of the rare, mint-condition G.I.Joe items you'll see in these pages. He has the best personal collection of these items that I've ever seen east of the Rio Grande. (He also owns a Dr. Evil in a dress.) Paul Levitt, the FX Channel's Japanese Monster Maven and chemist extraordinaire, also tossed in his Dr. Evil and a Colorforms alien or two.

As far as the west coast, I must thank Neil LaSala, one of the true nice guys in G.I.Joe collecting, for providing shots of the dressed dolls in his collection. Thanks to Roddy Garcia (that's pronounced Roady, by the way) for supplying shots of some of his ultra-rare boxed sets and Joes. Roddy is one of the pioneer G.I.Joe collectors who first drew Hasbro's attention to the vast adult interest in its most famous product.

Special thanks go to Earl Shores and Chuck Eckles who, between them, have virtually cornered the market in Johnny Hero expertise. I'll bet half of the readers of this book will never have heard of Mr. Hero until today. Earl and Chuck have been keepers of the Johnny flame.

Since I was in diapers through most of the sixties, I couldn't rely on first-hand knowledge in this book as much as in my others. This time I occasionally had to do actual research (ack) and I must gratefully acknowledge the work of experts like John McGonagle, Bill Bruegman, John Michlig, Rick Polizzi, Bill Sikora and T.N. Tumbusch, Lenny Lee, Jim Main, Don Cober, Gregory Boquist, James Tomlinson, Thomas Wheeler, James DeSimone, and many others whose names I didn't have a chance to look up. Their articles and books have educated me in the rarities of the toy aisles that existed while I was still merely an awesome infant.

A special shout out to Dave, Rosemary, Lorraine, and all my Columbus Farmer's Market pals.

Publisher Peter Schiffer and Editor/Photographer Jeff Snyder will expect to see their names here, so I better mention them. After all, I'm sure they did something. Oh yeah, Jeff took most of the pictures. Also I want to thank Blair Loughrey for the breathtaking design work he has done on my books. If he's assigned this one as well, we can all sit back and enjoy.

I want to thank the angelic Myrka Dellanos-Loynaz, sultry Mercedes Soler, and incomparable Maria Celeste Arrarás for the way their Spanish-language news magazine program *PRIMER IMPACTO* brought me back to the real world every evening. After a ten-hour work day thinking about bizarre stories, strange characters, and alien invaders, there's nothing like a dose of reality—Univision style!

And it certainly can't hurt to thank God for my writing talent and the sensibility to use my powers for the cause of good, and never evil. Well, almost never.

And, of course, this book is dedicated to my parents, John and Janice Marshall, for believing in me and providing me with that first G.I.Joe at Christmas 1971. It started me on a lifelong love of action figures—a love that even cranking out three books hasn't been able to diminish.

Introduction

For me, the Sixties mean three things: Cold War spy adventures, wacky outer space TV shows, and the arrival of mankind's greatest gift. Me!

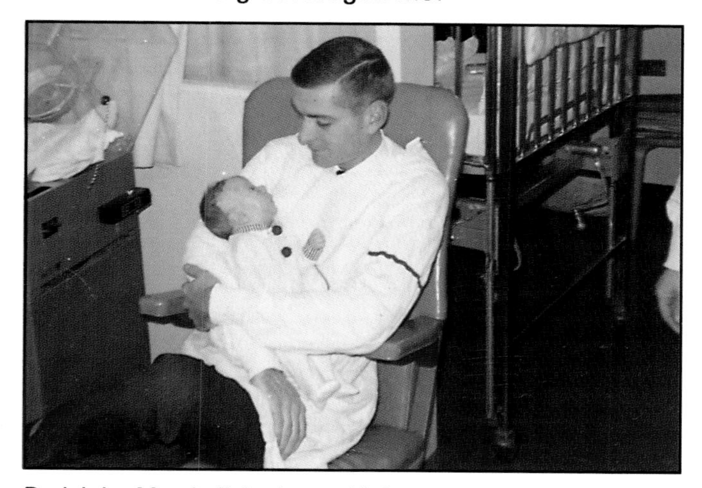

Dad John Marshall Junior and infant author John Marshall 3rd in Mount Sinai Hospital. *Courtesy of Marshall Archives.*

I arrived in this world a little after nine o'clock on the evening of October 2, 1964, in suburban New York. I was switched to a Manhattan hospital because I had a life-threatening birth defect. I had a condition called choanal atresia, in which my nasal passages were closed off by cartilage and I had to breathe through my mouth—yet somehow still eat and all that stuff. Well, to make a long story short, after a few months in one of those display case things at the hospital, I was able to come home. As an adult, I can't pass a display case without looking at what's inside it. Empathy, I guess.

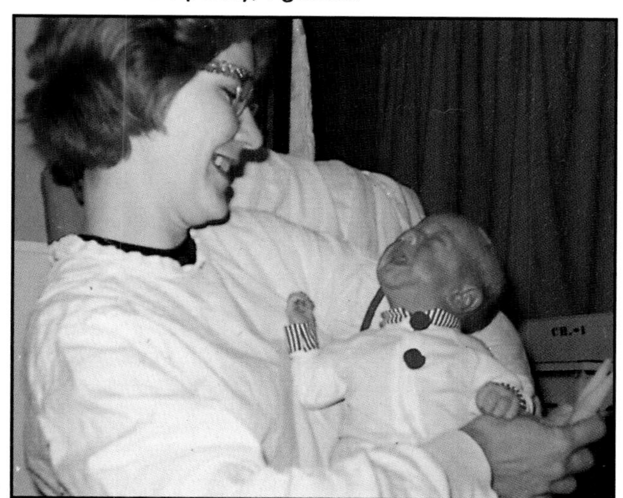

A glowing red John Marshall literally lights up the room. *Courtesy of Marshall Archives.*

It wasn't easy even at home, because I needed an airway in my mouth to keep it open at all times. Air is a staple in the health of many babies. Nevertheless, my parents took turns feeding me, changing, me, etc., always watchful that my little mouth was open. (Needless to say, I was never allowed to develop the habit of sucking my thumb. It would have killed me.)

It was not all peaches and cream, but after a few touch-and-go months, I was old enough to have the presence of mind to keep my mouth open, and I have spent a lifetime keeping my mouth open ever since.

A few years passed and we moved from Babylon, Long Island, to a nifty old, bat-infested house in Lambertville, New Jersey. As you East Coast readers know, Lambertville has since become a mecca for antiques enthusiasts and toy collectors. A coincidence? We can only wonder. Maybe Jonathan Frakes will host a TV show about it. Hopefully not, though.

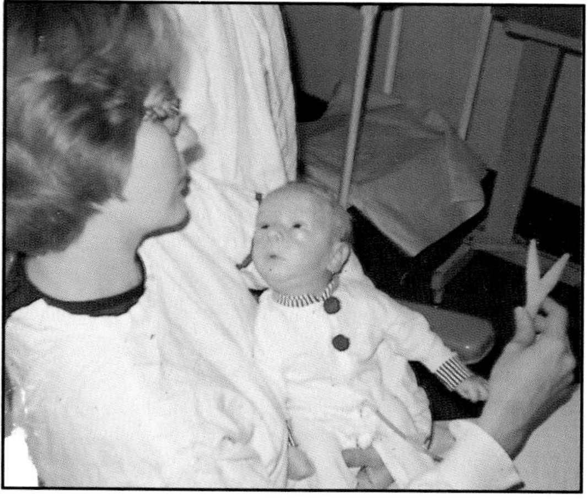

TV host Cristina Saralegui interviews an alien? No, it's a more subdued author-cito and attentive Mommy. *Courtesy of Marshall Archives.*

Our next move was to a sunny cul-de-sac in Willingboro, New Jersey, home of Olympic champion Carl Lewis, the Johnny Hero of his day. It was in Willingboro that, from ages four to six, I developed some of my earliest memories that make for good space filler in my books. Like this one, for instance: I remember my dad on the patio reading a paperback of the James Bond spy novel *You Only Live Twice*. The cover had a picture of a balloon with a skull face painted on it that repeatedly freaked me out. I kept my distance from that book at all times, and if I had to be in the same room with it, I turned it face down.

I also remember being in a kid gang, just like *The Little Rascals*. One time we peed on a guy's front lawn and he chased us all back to our respective houses. That day I created my first really well-performed B.S. routine. I said I'd been there, but hadn't participated in the piddling proceedings—and believe it or not, I fooled everybody!

Our next-door neighbors put on a big spook show every Halloween. Not only did they do up the front lawn, they performed in costume and did a complete horror show. It was awesome. Remember, this was in the days when TV monster movies were normally hosted by guys dressed up as vampires or mad doctors. Our local guy was Philadelphia's own Dr. Shock. We also had a weekday afternoon host named Wee Willie Webber who would do wraparounds between Spider-Man, Marine Boy, Speed Racer, and Ultraman. I remember that when we moved into our last house, outside of Mount Holly, Wee Willie assured me (over the airwaves) that even though I was in a new town, they still got Channel 17, and Ultraman was still on in the afternoons. (I distinctly remember him crossing his arms, mimicking Ultraman's firing position and saying, "And then, after Speed Racer, Ultraman!") Boy was I relieved.

The Seaview dives to new depths of adventure in this Remco playset. *Copyright 1965 Sears, Roebuck and Co.*

Since I was so young, most of the really cool 1960s stuff was actually "discovered" by me in reruns in the early 1970s. The only 1960s TV memory I have is of watching the 1969 moon landing, the one where Neil Armstrong says "This is one small step for a man, one giant leap for Mankind." But the radio transmission cut out the word "a" and for thirty years now people have been saying he made a syntactical mistake. Give the guy a break. He walked on the Moon, fer cryin' out loud. I also do not support the politically correct agenda to overdub "Mankind" with "Humankind" on all recordings of the event. Let's see a girl walk on the moon first!

Anyway, one of the few shows I had discovered on TV when it was actually originally on was *Lost In Space*, which premiered right around my first birthday. In fact, when the moon landing was being broadcast, my parents said to me, "How about that, Jay? A man's walking on the moon!" To which I replied, "Big deal! They do it on *Lost In Space* all the time!" True story.

And so (it must be pretty clear to you by now), I didn't really get around to playing with action figures during the 1960s. I made up for it later, though.

"Don't get any ideas, Krypto," says Captain Action to his space mutt. *Courtesy of The Toyrareum, Ocean City, New Jersey.*

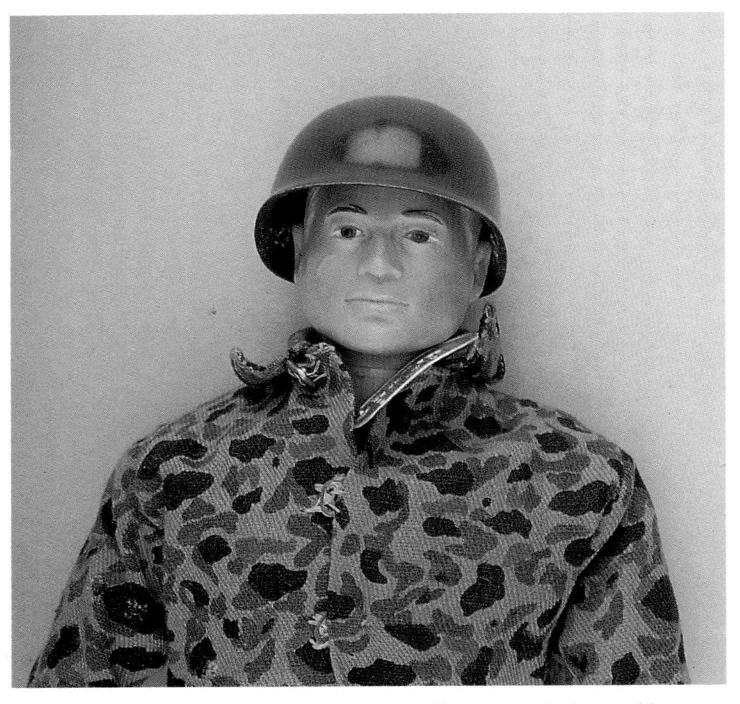

This 1960s G.I.Joe Marine is showing off an army helmet. Although in good supply in the 1960s, nice condition helmets like this one are hard to come by today. *Courtesy of Fred Mahn.*

The Panther Jet was made by Irwin for G.I.Joe under license from Hasbro. This fellow has lost his front wheels, a common problem. *Courtesy of The Toyrareum, Ocean City, New Jersey.*

"Howdy Ma'am," says Johnny West. *Courtesy of Play With This.*

Look out! A Russkie! Where's the jet when you need it? *Courtesy Of Play With This.*

The actual story of the history of action figures in the 1960s is the story of the development of an entire product category. G.I.Joe was the first, and, hard as it may be to believe today, it was pretty tough convincing buyers at the 1964 Toy Fair to take a chance on a doll—a dress up doll—for BOYS. That's why Hasbro's marketing goons were strictly instructed to sell G.I.Joe as a soldier. A fully-poseable figure who was ready for action...an *action figure.*

It was, in fact, a huge hit for everyone who picked it up. The rest is history. Marx, the venerable playset manufacturer and no stranger to little men designed for kids to play with, created their own action figure lines. These included a soldier (Stony Smith), Vikings, knights, and the popular Johnny West line. They went for an all-plastic approach, creating fully-dressed, sculpted statues. Their joints were cut right into the outline of the figures, but disguised in all the natural break points of the human body (hips, elbows, knees, etc.). Eventually, as clothing for action figures became more expensive to manufacture, producing fully-sculpted "dressed" characters became standard practice in the 1980s and 1990s. Only in the last few years have we seen the return of mannequin-style dress-up action figures, and that is a direct result of adult interest in action figure collecting.

Licensing played a key part in the early days of action figures. Gilbert, the erector set giant, snapped up the licenses for James Bond, *The Man From U.N.C.L.E.*, and *Honey West*, a TV series starring Anne Francis (of *Forbidden Planet* fame) as "TV's Private Eye-Full." Ouch! They also produced their own astronaut figure called Moon McDare. No, seriously! But although the figures were the same size as G.I.Joe, and had finely-detailed accessories, their bodies were stiff and virtually unposeable, tremendously reducing their play value. This is the only possible reason why these lines (especially James Bond) did not flourish.

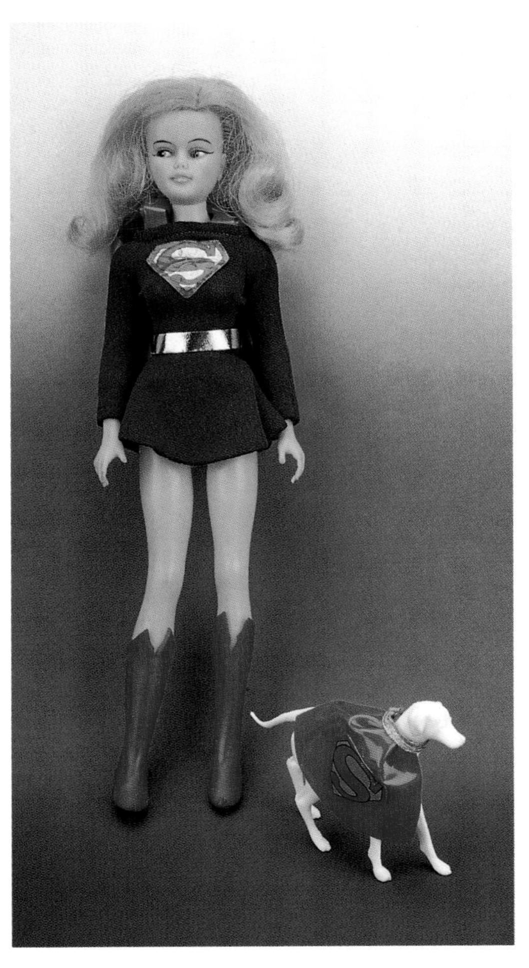

Krypto's at it again, this time with Supergirl from Ideal's rare Super Heroines line. *Courtesy Of The Toyrareum, Ocean City, New Jersey.*

Here's a typical Sears catalog page showing some 1966 spy goodies, Stony Smith, and some tank that's not an action figure. *Copyright 1966 Sears, Roebuck and Co.*

A cool color Sears page showing Captain Action from our Captain Action chapter, and Ideal's super heroes from the chapter titled Super Freakout Grab Bag! *Copyright 1966 Sears, Roebuck and Co.*

The next craze after spies was super heroes, as popularized by the awesome (and ageless) *Batman* TV series. Ideal tried their hand at an action figure, producing a 12" super hero called Captain Action who was almost as poseable as G.I.Joe. Plus, Cap had the improvement of flexible, natural-looking hands. His gimmick was that he could be dressed in Halloween-style costumes of many (indeed, most) of the most popular comic book and comic strip heroes of the day. Unfortunately for people who had waited their lives for, say, a G.I.Joe style Superman, when Captain Action was dressed in a Superman costume he looked exactly like just that very thing—a guy dressed in a Superman costume.

Fortunately, Ideal also released smaller, toy soldier size figures of the super heroes from DC comics. These were available in various sets and combinations. The good guys included Superman, Batman, Robin, Flash, Aquaman, and Wonder Woman. Each could be bought in painted or unpainted versions. Villains for the line, all available unpainted, included The Joker, a Wonder Woman villain called Mouse Man, two Justice League villains named Brain Storm and the Key, and the heroic Thunderbolt of Johnny Thunder fame posing as a villain. Plus, there were two generic menaces: a two-headed dragon named Koltar and a rampaging robot.

Ideal didn't have a monopoly on cool playsets. Remco released a *Voyage to the Bottom of the Sea* playset, complete with Seaview sub. Mattel added a *Lost In Space* set to its popular "Switch 'N Go" line. It featured the only vintage play figures of the Robinsons, Don West, Dr. Smith, The Robot, and even Debbie the Bloop, plus a Chariot and a Styrofoam Jupiter 2 spacecraft. Astonishingly, none of Irwin Allen's TV series became fodder for proper action figure lines, although their various hardware is all well represented in Aurora's model kits—except, oddly, for the conspicuously-absent Jupiter 2!

Other TV licenses included a line of Marx *Rat Patrol* figures (complete with Jeep) and the very Marx-like *Bonanza* figures from American Character. These included Ben, Hoss, and Little Joe. Parnell Roberts left the series but his character was released as The Outlaw.

A similar thing happened over at Marx. Their prototype James West figure (resembling Robert Conrad) had his body changed to Sam Cobra's and his head altered into Captain Maddox, both in the Johnny West line. Marx had better luck with Stephanie Powers as *The Girl From U.N.C.L.E.* Her doll made it all the way to store shelves—although most of those shelves were in Europe, not America.

If you don't know him now, you soon will. He's Johnny Hero. *Courtesy Of Earl Shores.*

The last major licensing event for action figures in that first decade was when major league football and baseball granted permission to Rosko Industries to produce official team uniforms for the short-lived 13" Johnny Hero doll. Like Captain Action, Johnny Hero could become anybody you wanted. But where Captain Action had a limited range of major characters, almost all the existing major league teams at the time had representative uniforms for Johnny to wear.

The final major trend in action figures was inspired not by TV or comic books but by the real heroes of the space race. Mattel created an entirely new type of action figure in Major Matt Mason, a six-inch astronaut constructed of sculpted rubber. He was all one piece—spacesuit and all—and built over a wire frame so his limbs could be bent into position. His helmet was removeable and had a working visor. Although there were sci-fi elements in the line, the essential thrust of Major Matt was nuts-and-bolts space exploration. To that end, he was armed with a near-endless supply of gadgets which had actual working features.

Major Matt Mason was so popular, in fact, that he inspired two me-too lines from major companies. Colorforms issued its Outer Space Men, a group of mostly-gruesome aliens to scale with, and constructed the same way as, Major Matt and his pals. Also, Ideal produced a series of battery-operated robots called Zeroids that looked right at home in anyone's Major Matt scenario.

By the close of the decade, manufacturers no longer looked upon action figures as mere dolls for boys. They saw them now as an exciting and highly exploitable new product category.

The rules laid down by G.I.Joe and Johnny West gave a new generation of toys a wide field in the 1970s, a field in which I played, long and hard.

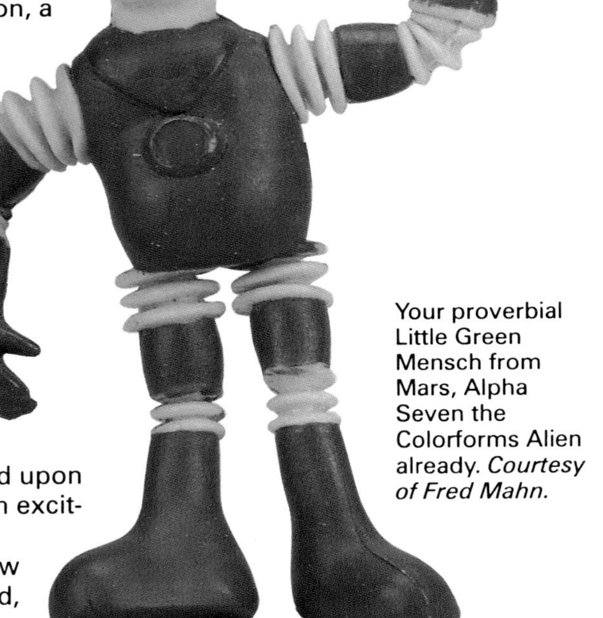

Your proverbial Little Green Mensch from Mars, Alpha Seven the Colorforms Alien already. *Courtesy of Fred Mahn.*

How To Use This Book

Now pay attention, 007:

Essentially, this is a guide to all action figures and related accessories released from the arrival of G.I.Joe in mid-1964 up to December 1969. Lines that continued into the early 1970s, like Major Matt Mason, are listed in total.

It's not always easy deciding what should and shouldn't go into a book like this. I had to seriously consider including the figurines made by Hartland, for example. Hartland manufactured a number of figures based on the famous cowboys of movies, TV, and history, as well as sports figures and other items. These predate the development of G.I.Joe (the first "proper" action figure, as M.C. Hammer might say) by several years. And yet, in my opinion, they are really more intended for display than play, much like model kits. They also are, in my opinion, products of a previous era. So they are not included.

Similarly, I have avoided, for the most part, listing figural model kits. Models really are a separate category, even though they were marketed in much the same way as action figures, and were bought by the same kids who bought action figures. But a model kit is not really a toy. I have bent the rules a bit in the chapters where figural merchandising is relatively low in proportion to the popularity of the characters. Thus, Aurora's, Revell's, and Airfix's figural models of the *Lost In Space* Robot, the cast of *Bonanza*, and James Bond vs Oddjob, respectively, are included. Aurora's extensive super hero and monster model lines aren't. There are several outstanding books on the market which cover figural models better than I ever could.

The chapters in this book are broken down according to the groups I have divided the items up into—in my head. You may argue that, because the Zeroids were made to be in scale with Major Matt Mason, that Zeroids deserve to be in his chapter, like the Colorforms Outer Space Men are. But the Outer Space Men are included with Matt Mason because they are made in the same way and are often confused with Matt Mason characters. Zeroids are robots and can stand alone, something even Major Matt can't do, literally speaking. That's what I say. And what I say goes.

Another thing you may not like is my definition of an "action figure," i.e., my determination of what does and what does not belong in this book. Generally, an action figure has to be a three-dimensional representation of a character, designed to interact with other toys of the same kind or similar kinds, in an adventure-related, creative play scenario. Sometimes I'll break the rules one way or the other and include non-articulated play figures (like those in the *Lost In Space* Switch-N-Go set) or figural play toys that aren't action figures (like the *Lost In Space* Robot by Remco). Does that make sense to you? If it does, you will enjoy this book immensely. If it doesn't, you will enjoy this book immensely anyway.

Generally speaking, this book only includes lines that were marketed to Americans and available in most retail stores around the country. This excludes, for the most part, foreign action figure lines related to American lines from the '60s, such as Britain's Action Man, the cousin of G.I.Joe. Some day I would like to do a book on these, but there are a few more years of research required to do an accurate book. However, important foreign additions to existing American lines, like the England-only accessory sets for Gilbert's James Bond doll, are included.

And speaking of "dolls," what of the terminology used in this book? Well, I often use "doll" and "action figure" interchangeably when I'm talking about poseable figures with removeable clothing. Some male-type men object to the objects of their fascination being referred to as dolls, but that's even sillier than Trekkies demanding to be called "Trekkers," like that has any less embarrassing a connotation. G.I.Joe is a doll. In fact, it is *legally* a doll. A few years ago, an imports issue came up in the 1990s over whether Hasbro, in reference to G.I.Joe, was bringing dolls or some other type of figure into the country. Our court system ruled that G.I.Joe was officially a doll. So there you have it.

Other terms you will see repeated *ad nauseum* in this voluminous volume include L.M.C., M.I.P., and "toy soldier-style."

LMC: This is the first of two headings in the price guides. It stands for Loose, Mint, Complete, and indicates that the item in question has all its original parts and is in more-or-less like-new condition. It can never be over-stressed how important completeness and condition are. Are you one of those poor, deluded souls who has had a capeless, beltless Mego Batman on your table at every antique show for years at $40? Are you wondering why those idiot collectors never buy it when it "books" at sixty or more? Well, remember the Mantra: Condition, completeness! Condition, completeness! Condition, completeness...

MIP: This is easy, it stands for Mint In Package. In John Marshall parlance, this means an item that is complete with all original packaging and paperwork. The outer package should be in something close to like-new condition, and the interior items should be in virtually straight-from-the-factory condition. The package may have been opened *if* the packaging can be opened and closed without damaging it.

You'll also see the term "toy soldier-style" a lot in this book. That generally means a figure that is around 3 inches tall and sculpted as a solid piece, like a statue.

Estimating value: The value of something in less than mint condition varies from buyer to buyer and seller to seller. It also really varies from item to item, especially when it comes to 1960s stuff. I recommend that for something in slightly played with condition (with most of its parts), you should expect to pay or get 60%-75% of the Loose, Mint, Complete price. On the other end of the scale, for especially rare pieces that are still factory sealed in crisp, shiny, like-new packaging, expect to pay or get an extra 25% or 30% above my Mint In Package listings.

How accurate is the information? Pretty darn accurate. But I do have a proviso for you to look at, and it's a long one. Unlike the following two decades, which I have already done books on, the period of the 1960s was shadowed in the early, misty world of experimentalism. In other words, the likelihood is far higher this time around that there might be variations in the design or packaging of existing items that I may simply not know about. There may even be a few items out there that qualify for inclusion in this book that I don't know about, none of my associates knows about—heck, maybe nobody knows about them anymore. I know that I come across as a big shot know-it-all, because, let's face it, I am. But that doesn't mean that I claim to know everything. After all, if you knew everything, as Doctor Who once said, where would the fun be? (A Dr. Who geek who makes fun of Trekkies, that's me.)

Which brings us to the prices given for items in this book. How did I arrive at the prices? How accurate are they? Well, they're based on my intimate knowledge of the field of action figure collectibles. I am constantly reviewing market trends and developments, I go to shows, and I read most of the toy-market publications. Simply put, it's my job to know the relative value of any item in this book. However, for this volume I reviewed prices with knowledgeable dealers, since values on these 1960s pieces fluctuate wildly, and auction results yield higher and higher values every few months. The prices listed herein are meant to provide a *general range* of what you can expect to buy or sell the items for. There's a worn-out old saying in the field of collectibles, that the true value of an item is the price a buyer is willing to pay for it. Don't blame me if the only Dr. Evil lab set you find costs a thousand dollars more than the price in my book. It may be the only one left in existence, or at least the only one in circulation. If you want it, get it. Got it? Good.

On the other hand, you can safely assume that you are getting a very good deal if you buy a boxed G.I.Joe Nurse at a toy show for $500 and you know that I have given it a much higher value. That, my friend, is what price guides are for.

You may notice a seeming disparity in the pricing; sometimes an item will be worth twice as much mint in package (MIP) as it is loose (no packaging), mint, and complete (LMC). Other times there won't be a huge difference between the loose and packaged prices. The reason for this is that I am incompetent. Just kidding. Really, it's because some things are rare enough that they are almost as hard to obtain loose as they are to obtain in the package (The Gilbert Sears-only James Bond in tuxedo comes to mind). Others, like many basic G.I.Joe accessories, are just the opposite.

Some listings make no logical sense upon first viewing. For instance, if a Captain Action Directional Communicator Set is worth $75 loose, mint, complete, why do the values of the individual pieces add up to over $75? The answer is really a question of taste. Y'see, several of the pieces are much harder to find, and are much more desirable, than other pieces in the same set. Sometimes the whole set complete is not worth a lot more than the key pieces are individually, those pieces that people really want. (For example, everyone wants a Captain Action Power Jet Pack but nobody cares about the boots and helmet.) That is why you have to trust me—it's my job to know these things.

Okay, I think that's everything. I hope you enjoy the book.

John Marshall, Valentine's Day 1998

Chapter One

G.I.Joe: America's Moveable Fighting Man

It was a dark and stormy night.

Well, almost.

Actually, it was a cold, rainy day in 1963. Our hero, Don Levine, head of research and development for Hassenfeld Brothers (Hasbro), trudged through a chilly downpour amidst the concrete monoliths of New York City.

A licensor named Stan Weston had been bugging Don about an idea for a doll, a doll intended for boys that would be marketed as a soldier. The doll would be based on a license Weston had the rights to, Gene Roddenberry's TV show *The Lieutenant*, a military-themed drama. But Levine had seen what was on the TV screen and turned green. The show was far too cerebral for any kind of kid's marketing campaign. (A problem that would plague Roddenberry on his next show, too.)

Action Soldier with black hair. *Courtesy of Play With This.*

Stan Weston, however, still wanted to push the idea of a soldier doll for boys. Levine was intrigued—how could such a thing be marketed? Would it sell? Would boys play with dolls?

As Levine shoved and jostled his way through crowds of soaked New Yorkers on Fifth Avenue, he passed an artist supply shop. In the display window was an artist's mannequin, a wooden reproduction of the human body, complete with many different joints, which artists use as a model when painting or sculpting human figures.

Don Levine halted in front of the window, transfixed. As speckles of raindrops spattered against the window, the image of the mannequin became obscured. Levine looked in closer, scrutinizing the figure. He began to see it in army fatigues, holding a small rifle scaled down the same way the figure was. A figure that came complete with little boots, little dog tags.

"It could work," a small voice said in the back of his mind. "It would be the ultimate toy soldier. Stand him up. Sit him down. Change his uniform."

"It could work," the voice insisted. "It really could."

Of course, this is just a dramatization, but nevertheless, this key event really did take place, according to Levine's own account of the history of G.I.Joe.

Merrill Hassenfeld was the second-generation president of the Hassenfeld Brothers Toy Company (Hasbro) and he knew his business both inside and out. He had helmed his company as its toy division grew in the 1950s and early 1960s, becoming the producer of Mister Potato Head and other popular products. But when he returned from a trip abroad and was hit with Don Levine's proposal, he had his doubts ... serious doubts. Mass-production of Levine and Weston's soldier doll would have to be done in the Far East, a big, big step even for a multi-million dollar company like Hasbro.

To insure maximum marketability, Don Levine had worked out an entire series of figures, with each character representing the four services. There was Skip the navy frogman, Rocky the marine, Ace the pilot, and a soldier figure whose name is lost to the ages. Hasbro's ad agency at the time knew that such a line had to be focused—all those different characters would only divide up the concentration of the marketing department. The line needed one single, all-encompassing name for kids everywhere to memorize and scream for. Inspired by the war film *The Story Of G.I.Joe*, Hasbro chose the generic name "G.I.Joe" for the embryonic product line.

Plus, to trademark the doll, it had to be given a special feature since you can't trademark the human body. (Of course, people try to trademark all sorts of things. Marvel Comics tried to trademark the word "mutant." I myself, in a mad scheme for world domination via communications control, tried to trademark the words *a*, *an* and *the*. But they kicked me out of the office.)

Anyway, to give G.I.Joe a distinctive, trademark-able look, a scar was sculpted onto his cheek. His thumbnail was also etched into the bottom of his thumb, not the top.

The prototype doll had 21 moveable parts and a patent was eventually filed for him on October 11, 1966. That's why early G.I.Joes have "patent pending" stamped on their butts. Later Joes, manufactured from 1967 on, are stamped with the best-known patent number in the world, 3, 277, 602. This patent lasted until the mid-1970s, when the newly redesigned "Life Like" Joe body was created.

Okay, so by this time, the initial hurdles were worked out. The toy obviously was a big risk, but its potential was just as obvious to most of the decision makers at Hasbro. But there was another big problem right at the start.

Toy buyers for the major retail chains would not want to wait until 1965 for a products solicited at the 1964 Toy Fair, so Hassenfeld okayed the production in mid-1963. He then selected a few choice buyers to come in for a special showing. So, even at its inception, G.I.Joe was an exception to the rules!

And so it began. Some buyers were crazy for it, and ordered in sufficient quantities. The other buyers, the *dumb* buyers, the ones who saw this doll for boys as some abominable freak of manufacturing, ordered lightly or not at all.

By the summer of 1964, it was clear that the pro-doll faction was right. G.I.Joe sold like crazy. The following October I was born and those two events formally ushered in *The Age Of Action Figures*.

G.I.Joe rapidly became a fixture in toy stores across the nation, but not just in toy stores. G.I.Joe was huge; drug stores, hardware stores, and any non-specialty retailer might conceivably carry him. Everyone wanted to get on the G.I.Joe bandwagon. The secret of Joe's success was the variety of ways in which Joe items were packaged. It broke down rather neatly as a three-tiered system. The dolls and the larger outfit sets sold for $4-$5, equipment sets in small boxes sold for $2-$3, and accessories on small, cellophane-wrapped cards sold for about a buck. This way, even small stores could arrange for a display rack of Joe accessories to bring in more kid traffic.

Aesthetically, the G.I.Joe line was divided into four services, with all the initial offerings falling into categories for either Action Soldier, Action Sailor, Action Marine, or Action Pilot. Craftily, Hasbro issued many of the same items in different packaging so that you might pay more than once for certain pieces. At the same time, many outfits could not be completed with just one purchase, but instead required piecing together. In fact, even the deluxe outfit sets were usually missing some key piece. Here's an example. After shelling out a whopping $4.99 for an M.P. uniform, you still had to shell out for the carded helmet that went with it—and the extra belt, holster, pistol, and billy club that came with it on the accessory card. It was hard to fault Hasbro, though. The pieces that were duplicated, usually, were the small ones that were often lost. If they weren't lost, well, you can never have too many gun belts and pistols, right?

Let's take an overview of what was available in the early years of 1964 and 1965. These basics formed the backbone of the 1960s G.I.Joe line. These are the elements that proved to be such big hits and fox-holed G.I.Joe in the hearts of kids forever.

The Action Soldier was properly attired to take on any kind of land mission. His optional clothing (i.e., boxed or carded stuff) included a marvelous green field jacket with a working zipper. And, unlike the G.I.Joes of today, this zipper was actually in scale with the rest of him, making for a very realistic outfit. This could be augmented by his Command Post Poncho in waterproof green, and accessorized with a lovely Field Pack with real brass buckles. After a hard day in the trenches, Joe could retire to his Pup Tent. It could actually be strung up like the real thing—for Real Tent Action!

Of course, the well-equipped soldier must accessorize; available to G.I.Joe were both the M-1 and carbine rifles, along with an ammo belt, bayonet, a canteen with a cloth cover, a shovel with a cloth cover, and even a full mess kit with knife, fork, and spoon. There were also a tripod mounted machine gun, an ammo box, and sandbags. Hasbro wisely sold the sandbags in carded sets so you could stack them as high as your imagination and budget allowed.

There were deluxe uniforms as well, which required the aforementioned multiple purchases to complete. But it was worth it. The Military Police set, originally issued in a kind of greenish brown, was composed of jacket, pants, scarf, helmet, and accessories, everything you'd

Action Soldier with blonde hair.
Courtesy of Fred Mahn.

This catalog page features the 1965 G.I.Joe line from Sears. *Copyright 1965 Sears, Roebuck, and Co.*

need. And fancier still was the range of snow gear available for Joe. This consisted of two main sets. The deluxe outfit, Ski Patrol, came with a hooded jacket, pants, skis, ski poles, goggles, and a rifle. You could accessorize with Mountain Troops: white backpack, snowshoes, pickax, and rope. The helmet was, of course, sold on a separate accessory card. Then all you had to do was pour detergent soap flakes all over the place to create a winter scene any time of the year! That's what Andy and George did.

Ahh, yes, Andy and George. Throughout the 1960s, A & G appeared in black and white, full page ads in DC comics. Drawn by Irv Novick, these ads told a story in comic book panels of Andy and George setting up their G.I.Joes. These vignettes clearly showed the appearance and function of the particular items advertised in each ad. Andy and George were real salesmen who could shill like nobody's business. "Gee, this new Mountain Troops backpack is white—just like the real thing!" Andy would declare. "Let's ask Dad to buy us a dozen more!" George would reply. (Well, I'm paraphrasing here, but that was the gist of the ads.)

The Action Soldier's pal, the Action Marine, shared most of the items with his comrade-in-khaki; but, some of these pieces were done up in camouflage for the marine, such as the helmet and the pup tent. Other, completely different items made for the Action Marine included a non-working paratrooper parachute pack and a flame thrower. Looking back, all things considered, having something as gruesome as a flame thrower accessory was really pushing it, but I guess it seemed appropriate at the time.

However ol' Jar Head really got the better of his dog face pal with two deluxe sets all his own. One was the Marine Parade Dress uniform, perhaps the most-copied, often imitated of all G.I.Joe uniforms. But better still were the items in the Marine Medic series. These included—and remember all this was done in scale with a foot-tall figure—stethoscope, bandages with pins, plasma bottle with needle, splints, crutch, and stretcher. There was also a green helmet with a medic sticker.

In fact, helmets were fairly plentiful back then. It is a surprise how genuinely rare helmets are today, considering that helmets were issued in various packaging for army basics, marine basics, navy, M.P., medic ... and the list goes on. Yet, helmets are extremely hard to find today.

For its Navy and Pilot lines, Hasbro got a little more creative and released some things that were a lot fancier than the average fatigues and helmets. The Sea Rescue set featured a yellow life raft marked "G.I.Joe" and a tan paddle, the first vehicle for G.I.Joe. The Navy Attack set featured a replica navy life jacket and signal flags, but to complete the outfit you had to buy the blue helmet on an accessory card. Other carded sets unique to the Action Sailor were Dress Parade (just the accessories) and a machine gun set with a blue ammo box. But by far the most impressive outfit set was the Shore Patrol, with its navy blue traditional sailor suit, sea bag, and M.P. accessories. The earliest version had a zipper which ran from the bottom of the shirt all the way up one wrist. Later, it only ran up to the armpit and required more of a struggle to get Joe into it. Plenty of fake sailor shirts abound, but these have no zipper and are of lightweight material.

There was also a scuba suit which consisted of a real, waterproof black rubber top, pants and hood, and accessories including tanks, mask, knife, and depth gauge. The rubber, even under the best of circumstances, had about a 20-year lifespan, and the suits today are only found unrotted on the rarest occasions—even mint-in-box unopened specimens usually have at least some decomposition.

But the greatest early water-related set was the boxed Deep Sea Diver set. This came in an oversized box much larger than the regular uniform sets, and the diving suit itself came mounted on a dummy body. Unlike other deluxe outfits, this one came complete in one package but it was priced accordingly at around $5.99. The outfit was a near-perfect copy of the World War Two-era diving suit. It had a brass-colored, three piece helmet and chest plate assembly. The face plate on the helmet opened on a hinge. The suit itself was actually waterproof, with elastic bands to seal off the neck, wrists, and ankles. Joe had a belt with lead weights, and lead shoes to fit over his boots. An air tube was provided and, through lung power, Joe could actually be raised and lowered in the water.

The Action Pilot, at least initially, got fairly short-changed on sets. Most of his equipment duplicated everyone else's, albeit with the occasional color change. Two examples are the gray, vinyl-strapped field phone and larger blue backpack field phone. But there were a few exclusive pieces for the Pilot, such as a gorgeous gray Scramble jumpsuit that came with a clipboard and a Joe-scale working red pencil. There was also a carded white crash helmet to go with that suit. But best of all was the Air Force Dress Uniform, a blue affair with a shirt and tie, socks and shoes, and brass buttons on the pockets.

There were also some accessories that anyone could use. A set of flags featured Old Glory and flags for each of the four services. A footlocker was issued briefly in vinyl before the wood format became the model of legend. Like everything else, the quality was evident in the detailing: rope handles and real brass hinges.

Hasbro racked up $16.9 million in sales in the first half-year alone, and went on to do double that amount in 1965, backed up by an outstanding and invasive TV advertising campaign. New for 1965 was an Action Soldier molded in brown plastic and sold as a "Negro Soldier," even though he had the same head as the white Joes all had. Nevertheless, it was both financially and culturally smart to recognize African-Americans as the growing consumer force they had become. Personally, I classify myself as a Quaffian-American, because I'll quaff down any liquid that is put in front of me.

Also in 1965, we received the first, and best, large G.I.Joe vehicle of the 1960s, the Five-Star Jeep with its "moto-rev sound," working searchlight, and shell-firing cannon. This is probably the only real, true-to-scale and true-to-detail vehicle for Joe.

By 1966, as far as the sales results went, it looked like Hasbro had very little to worry about. Despite competition from Gilbert's nearly-immobile James Bond and *Man From U.N.C.L.E.* figures and Marx's finger-pinching Johnny West, Joe was tops in boys' toys.

And then, with its top position virtually secure, Hasbro laid out the competition permanently with the addition of an entire line of deluxe figures. This series represented the sum total of the ability of Hasbro's design and manufacturing teams.

The Green Beret was Hasbro's modern-army hero, straight from the rice paddies. His all-new uniform was a near-perfect copy of the real thing. His beret was sculpted rubber so that it held its perfect shape sitting on his head. His all-new M-16 rifle and shell-launching bazooka added real battle excitement. Geez-I-Joe! I sound like Andy and George!

But even bigger news was the Soldiers Of The World series. These figures had all-new head sculpts, a first for the line. Five of them had a kind of sourpuss expression and what is called a "Nordic Look." And they had no facial scars. The sixth figure, the Japanese Imperial Soldier, had a head sculpt uniquely his own.

At this time in 1966, G.I.Joe went international in another sense. Foreign countries were beginning to adopt G.I.Joe as their own. In Britain, he was called Action Man, and was imported with very few adaptations, except for the obvious, of course—the Green Beret replaced the British Commando in their "Soldiers Of The World" series. Eventually G.I.Joe would be produced for Italy, Spain, and many other countries in Europe, South America, and Japan.

But the biggest news in 1966 was that G.I.Joe had entered the Space Race. The all new Mercury Capsule came with a sliding hatch and a complete space uniform, including helmet and life support systems.

Deluxe uniform sets got better and better in 1966. The Green Beret uniform, with bazooka, was also available as a boxed outfit without a figure and was joined by some awesome new items. The LSO Deck Commander had a realistic striped uniform, signal paddles, and an easily lost cloth headpiece. The Crash Crew Fire Suit was a masterpiece in silver, right down to the "asbestos" lining on the palms of the gloves, and the miniature, working tools in the tool belt. No person who has ever handled those little pliers has failed to be moved. New accessories included a bunk bed and a mine detector that lit up when passed over any of three metal mines.

Why did Joe need a Deck Commander and a Crash Crew uniform? Why, to interact with his new JET PLANE! At long last, G.I.Joe was to have an extensive line of vehicles. The problem had always been that quality vehicles were expensive to produce. That's why only the very best items (i.e. the jeep and the space capsule) made it into production.

He yam what he yam, the redheaded Action Sailor. *Courtesy of Play With This.*

But Hasbro struck a deal with the Irwin toy company to produce low-cost vehicles in scale with G.I.Joe. These were not knockoffs, but licensed items manufactured by a separate company. This was nothing new. By 1966 G.I.Joe had become a marketing phenomenon and Hasbro had licensed him for Halloween costumes, electric drawing sets, child-size versions of G.I.Joe's equipment, and more. Some of these products Hasbro made and some they did not.

But this was the first time that Hasbro had allowed another company to manufacture items for G.I.Joe to use as part of the toy line proper.

These vehicles, mind you, were cheap compared to the genuine vehicles. While the Jeep and Space Capsule were made using the reliable injection-molding process, the Irwin vehicles were blow-molded, resulting in fairly fragile items. On a personal note, when I borrowed the Panther Jet from Toyrareum to photograph for this book, I basically bit my nails for the entire time that I had it in my possession, fearing that if I dropped a tissue on it, a wing would break off. Nothing went wrong, though.

Nevertheless, these vehicles were greatly prized back then, and are sought-after collectibles today. They weren't manufactured quite to scale, much like Irwin's Barbie vehicles, so Joe's head and shoulders had a tendency to stick out of the vehicles and you could almost hear him saying, Bugs Bunny style, "Ehh, I usually take a size toidy-six!"

In addition to the vehicles already mentioned, there was also a Personnel Carrier/Mine Sweeper and Amphibious Duck troop transporter, which were pretty large but not really in scale.

And then came 1967, the year that G.I.Joe hit its artistic peak for the 1960s The line was expanded dramatically, with not a hint of a drop in quality. The most striking of all the innovations was that G.I.Joe could now TALK! Talking versions of all four service figures arrived, equipped with a pull-string dog tag and eight commanding phrases.

This brown-haired Action Pilot seems very proud of his separately-available Scramble Jumpsuit. *Courtesy of Play With This.*

Some of the best (and shortest-lived) accessory sets came out in 1967. Heavy Weapons featured a flak vest and M-60 machine gun. The Weapons Rack included three different rifles and a grenade launcher. The Combat Engineer series included carded accessories like a surveyor's tripod and working jackhammer.

But by far the most desirable items from 1967 are the deluxe outfit sets. Two new versions of the MP uniform were released, one tan, one green. Tank Commander featured a faux leather jacket and "sheepskin" collar, plus a vinyl helmet equipped with goggles and microphone. Rarest of all was the green Marine Jungle Fighter outfit which came equipped with the same accessories as the Australian Jungle Fighter.

The best outfit of all was Joe's new Fighter Pilot outfit with its snazzy gold and black helmet, green G-pants, cloth life vest, and, at last, a real working parachute.

Hasbro experimented with some adventure-oriented sets as well. There was Sabotage, complete with a black night landing raft, "grease" gun, and real wool knit cap. Deep Freeze included a fur parka, supply sled, and cold weather gear. Breeches Buoy featured a life-line pulley and harness and a weatherproof yellow nylon rain suit.

But the most interesting of all were the new Cadet series outfits. There were three: Air Academy, Annapolis, and West Point—accurately detailed right down to the last brass button. Caps, sashes, socks and shoes were augmented by ceremonial swords and scabbards. If I had a dollar for every sash I've thrown out, thinking it was a curtain tie ...

Meanwhile Joe had a semi-new vehicle. Inspired by the *Rat Patrol* TV show, Hasbro reissued their jeep in tan. It was equipped with a mounted machine gun and a driver dressed in Australian-style desert clothes and orange, white-rimmed goggles. There was also a Sea Sled, available both with and without an accompanying frogman figure. The final major release was a Hasbro-made Crash Crew Fire Truck, a blue monster which held Joes in the front and the back, had an extension ladder, and a working water hose. These two were on every kid's holiday wish Christmas list in 1967.

Meanwhile, in Canada, Hasbro released a Canadian Mountie outfit. They also experimented with a non-military Action Joe line. This was a series of two driver-and-vehicle sets. One was a Race Car and Driver set and the other was a State Trooper and Motorcycle set. These items were never available in the States at the retail level.

Figuring correctly that they'd covered everything thoroughly, Hasbro decided to go where even they had never gone, and take the ultimate plunge.

They released a G.I.Joe ... *for girls*.

This figure, a Nurse nicknamed "Jane," was ten and a half inches tall and had 14 moveable joints. She came complete with a medic set, and her cloth accessories were alternately white or green. To equalize things, there was also a mail-order doctor uniform.

Alas, much like Lionel's pink train for girls, the Action Girl never found an audience. Boys wouldn't touch her and girls weren't keen on a combat-oriented figure. And so, Hasbro's one big Joe flop of the sixties came and went, zippedy-zap. Of course, it's just about the rarest item to locate today.

By the way, on a related note, G.I.Joe was actually scaled along the same lines as Barbie. They are exactly the same height, which means that in "real life" Joe would stand five-foot-nine, eye to eye with the Barbster.

And then, as 1968 dawned and Hasbro thought it could do no wrong (Action Nurse Jane notwithstanding), the bottom fell out of the war wagon. The war in Asia was dragging on, and parents were getting sick of the real war, the fake wars on TV shows and in movies, and, especially, the wars waged in their living rooms with G.I.Joe.

From the mass assault of new G.I.Joe product in 1967 there was an astonishing drop off in manufacturing in 1968. Hasbro scrambled to repackage and/

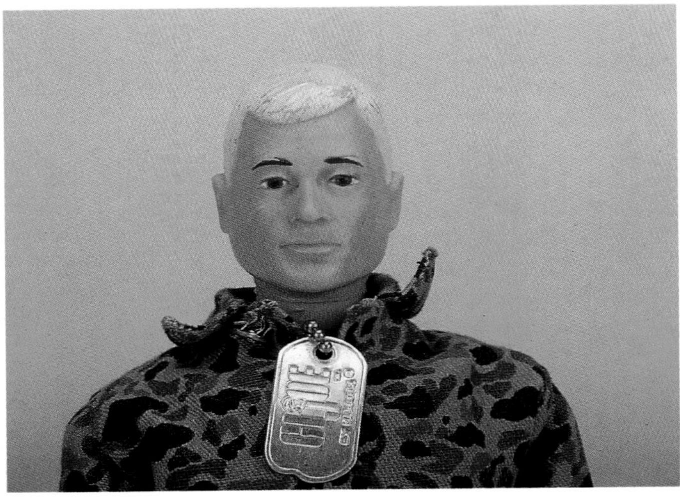

And here's a special guest star hair color—white? Apparently an error occurred at the factory and this figure was sprayed with the white base coat but no blonde coloration. *Courtesy of Fred Mahn.*

or otherwise jazz up what they had. They combined talking G.I.Joes with various equipment combinations. They starting putting outfit sets in big, display-oriented photo boxes showing real soldiers at work.

But nothing worked. G.I.Joe went from multi-page listings in the earlier Sears catalogs to barely a page in 1968. The end had finally come. It was time to do some rethinking.

G.I.Joe was back in full force the following year, but there had been big, BIG changes. The quality and attention to detail were both still there, but now G.I.Joe had a new focus. War was now no longer his main concern. Now he would be more of a soldier of fortune, seeking adventure around the world and in outer space.

This was the series called The Adventures Of G.I.Joe. It would reinterpret G.I.Joe for a new generation. The basic Joe was now an Adventurer, who wore a white T-shirt, tan cap, and tan pants, plus a brown shoulder holster and pistol. He came in both Caucasian and African-American versions. Additionally there was a black-sweatered Aquanaut for sea missions and a terrific Talking Astronaut who uttered real-life Mission Control phrases like "Ten seconds to lift-off, and counting!"

Three different equipment lockers were issued for the new team. There was a red Adventure Locker, blue Astro Locker, and green Aqua Locker.

The big developments equipment-wise were that the outfit sets were larger and all of them were complete in one package (no more multiple purchases required to assemble an outfit). Also, with the new adventure format, the old military-themed sets were gone. Instead, each equipment set was designed around a specific adventure. A comic book was enclosed with each set and all the major pieces of equipment were provided for the kid to act out the adventure. He was encouraged to create new endings and even create entire new adventures himself.

This theme would be further developed in the 1970s, when G.I.Joe would get a new look to go with his new image.

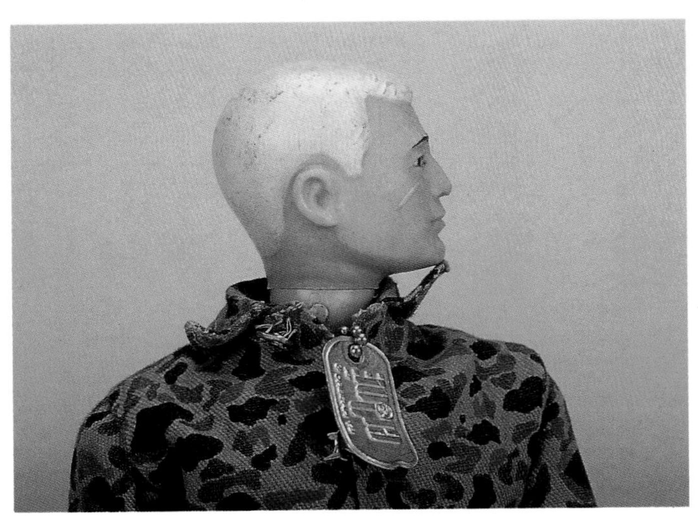

Here's a side view. Note slight traces cf blonde if you can. *Courtesy of Fred Mahn.*

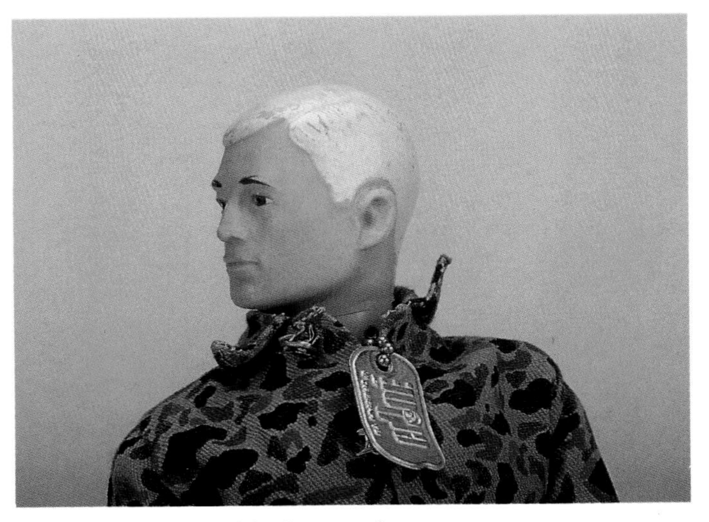

The other side. Hey it's that guy from Jonny Quest! *Courtesy of Fred Mahn.*

"De plane, Boss!"

Boxed Sets

Field Jacket Set	$50	$150
Field Pack Set	$50	$150
Bivouac Sleeping Bag Set	$50	$150
Bivouac Pup Tent Set	$60	$175
Sabotage Deluxe Set	$175	$950
Military Police Set (brown)	$125	$500
Military Police Set (green)	$225	$700
Military Police Set (tan)	$500	$1800
Ski Patrol Boxed Set	$75	$250
Snow Troops boxed set	$40	$175
Heavy Weapons	$125	$800
Green Beret boxed uniform only, no doll	$150	$700
West Point Cadet	$200	$750

"Boss! De Plane!"

"De plane!"

"De plane!" No nose landing gear, but I pointed that out already in the Introduction. *All courtesy of The Toyrareum, Ocean City, New Jersey.*

Action Sailor		
Carded Sailor Accessories, general range	$30-$40	$60-$80
Special Items		
Navy Attack Helmet	$35	$70
Shore Patrol Helmet	$45	$125
Life Ring	$45	$95
Boxed Sets		
Navy Frogman Underwater Demolition	$200	$600
Sea Rescue Boxed Set	$75	$250
Navy Attack	$60	$195
Landing Signal Officer	$125	$500
Deep Sea Diver	$200	$900
Deep Freeze	$195	$800
Breeches Buoy	$175	$800
Annapolis Cadet	$250	$950
Shore Patrol (full or half zipper)	$125	$425
Shore Patrol w/field radio	$400	$1500

Action Marine		
Carded Marine Accessories, general range	$20-$30	$40-$50
Special Items		
Communications Flag Set	$150	$225
Medic Accessories	$45	$125
Bunk Bed	$75	$195
Mortar Set	$65	$150
Weapons Rack	$85	$225
Boxed Sets		
BeachHead Assault Tent Set	$65	$225
BeachHead Assault Field Pack Set	$65	$225
Marine Medic Set	$95	$425
Communications Post Poncho	$175	$350
Combat Paratrooper	$65	$225
Dress Parade	$75	$225
Demolition (Mine Detector)	$65	$195
Tank Commander	$225	$1500
Marine Jungle Fighter	$350	$2500

Action Pilot		
Carded Pilot Accessories, general range	$30-$40	$50-$60
Special Items		
Scramble Crash Helmet	$40	$125
Scramble Parachute Pack	$35	$75
Air Police Helmet	$100	$225
Mae West Vest	$75	$150
Boxed Sets		
Survival Raft & Equipment	$125	$500
Scramble Flight Suit Set	$95	$350
Crash Crew	$95	$350
Air Cadet	$225	$850
Astronaut Suit	$95	$900
Air Sea Rescue	$195	$700
Fighter Pilot Suit w/Parachute	$395	$2200

This gorgeous page shows how, in 1966, Joe began to multitask. *Copyright 1966 Sears, Roebuck and Co.*

Above: The Astronaut, Capsule, and Sears Exclusive flotation collar. It's a floater. *Courtesy of WO1 Neil T. LaSala USMC.*

Left: Here's the Jeep, making its debut in the Sears catalog, Joe's first and best military vehicle. *Copyright 1965 Sears, Roebuck and Co.*

And here it is live. *Courtesy of WO1 Neil T, LaSala, USMC.*

Foreign Soldiers Accessory Cards

Japanese	$175	$395
Russian	$125	$295
German	$125	$295
British	$125	$295
Australian	$50	$295
French	$50	$295

(A large cache of French and Australian accessory sets in plain plastic bags turned up in a find in the 1980s)

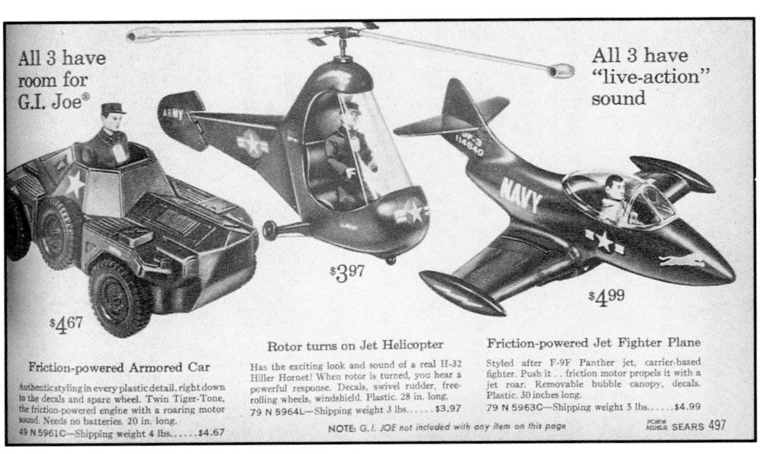

Three Irwin-made vehicles from 1965.
Copyright 1965 Sears, Roebuck and Co.

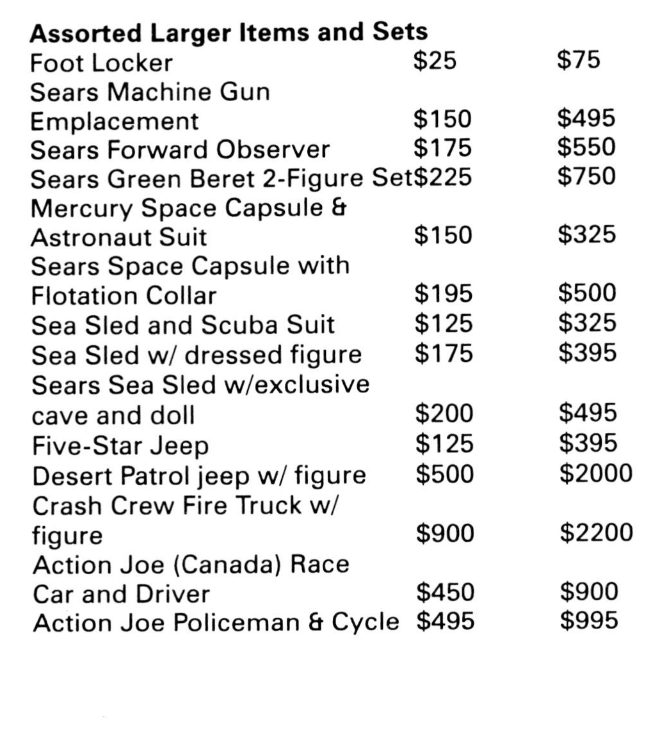

Assorted Larger Items and Sets

Foot Locker	$25	$75
Sears Machine Gun Emplacement	$150	$495
Sears Forward Observer	$175	$550
Sears Green Beret 2-Figure Set	$225	$750
Mercury Space Capsule & Astronaut Suit	$150	$325
Sears Space Capsule with Flotation Collar	$195	$500
Sea Sled and Scuba Suit	$125	$325
Sea Sled w/ dressed figure	$175	$395
Sears Sea Sled w/exclusive cave and doll	$200	$495
Five-Star Jeep	$125	$395
Desert Patrol jeep w/ figure	$500	$2000
Crash Crew Fire Truck w/ figure	$900	$2200
Action Joe (Canada) Race Car and Driver	$450	$900
Action Joe Policeman & Cycle	$495	$995

Above: We just wanted to show you this gorgeous, mint black-haired head.
Courtesy of Play With This.

Left: Shore Patrol, shore to please.
Courtesy of Play With This.

Irwin Vehicles

Motorcycle and Side Car $150 $295
Armored Car $125 $250
Staff Car $495 $950
Jet Plane $395 $650
Jet Helicopter $150 $395
Personnel Carrier/Minesweeper $275
$750
Amphibious Duck $275 $750

Two G.I. Joes of the
Green Beret
in this Jungle Outpost
hold off the enemy
with machine gun and
shell-firing bazooka

EXCLUSIVE SEARS SET $9.89

Now you can double the firepower of your forward outposts . . defend your headquarters from the enemy better than ever. Two G.I. Joes made of fully jointed plastic and standing 11½ inches high are dressed in full combat outfits (fatigue shirts and pants with boots, dog tags and green berets). They're ready to swing into action with a bazooka that actually fires one of six rocket shells. If enemy soldiers move in close, they'll be met with the machine gun, grenades or automatic rifle. Ammo box, cartridge belt, field telephone with earphone, camouflaged netting included. Plastic.
49 N 5978—Wt. 1 lb. 8 oz..Set $9.89

G.I. Joe reports enemy
positions with this
Forward Observer Set

EXCLUSIVE SEARS SET $4.97

Carefully hide military gear in this camouflaged observation post. Set includes G.I. Joe who's fully jointed plastic, takes any action pose . . 11½ inches high. His combat outfit is complete with fatigue shirt, pants, boots, dog tag, camouflaged helmet. Plus poncho, rifle, 4 grenades, field radio, telephone, wire roll, map case with maps, camouflaged netting, support posts. Buy it the easy way—order by phone. Shipping weight 1 lb. 5 oz.
49 N 5969.................Set $4.97

G.I. Joe Action-Soldiers of the World . . 11½-inch plastic, all realistically equipped

1 **French Resistance Fighter.** Underground hero in beret, sweater, pants and boots. With revolver, knife, grenades, radio set, submachine gun, more.
49 N 5992—Shipping weight 14 oz.....Set $4.97

2 **British Commando.** Proud hero in helmet, jacket, pants and boots. Also with gas mask, submachine gun, canteen, case, more.
49 N 5993—Shipping weight 14 oz.....Set $4.97

3 **Japanese Imperial Soldier.** Hand-picked defender in full uniform with gas mask, field pack. Also pistol, holster, belt, rifle, bayonet and more.
49 N 5990—Shipping weight 14 oz.....Set $4.97

4 **German Storm Trooper.** Uniformed with boots, helmet. Luger, machine pistol, grenades, holster, cartridge belt, field pack and more.
49 N 5989—Shipping weight 14 oz.....Set $4.97

5 **Russian Infantryman.** Expert at winter warfare in warm hat, boots, uniform. Also bipod machine gun, grenades, field glasses, more.
49 N 5991—Shipping wt. 14 oz.......Set $4.97

6 **Australian Jungle Fighter.** Uniformed bush warrior. Flame thrower, machete, entrenching tool, jungle knife, grenades and more.
49 N 5994—Shipping wt. 15 oz.......Set $4.97

SAVE THIS CATALOG . . you can order toys on pages 467 to 638 from now until August 1, 1967

Sears 511

Above: Here's Shore Patrol in the box. Wow! *Courtesy of Roddy M. Garcia, Houston, Texas.*

Left: This Sears page shows two Sears exclusive combo sets and all six Internationals! *Copyright 1966 Sears, Roebuck and Co.*

Below: A detail shot of the Foreigners. *Copyright 1966 Sears, Roebuck and Co.*

Adventures Of G.I.Joe Sets & Accessories

Adventure Locker	$75	$125
Aqua Locker	$100	$200
Astro Locker	$100	$200
Mysterious Explosion	$150	$495
Mouth Of Doom	$175	$595
Perilous Rescue	$225	$595
Danger Of The Depths	$150	$450
Secret Mission To Spy Island	$125	$450
Eight Ropes of Danger	$125	$595
Fantastic Freefall	$125	$595
Hidden Missile Discovery	$100	$495
Fight For Survival	$175	$695
The Shark's Surprise	$175	$595
Spacewalk Mystery	$175	$395

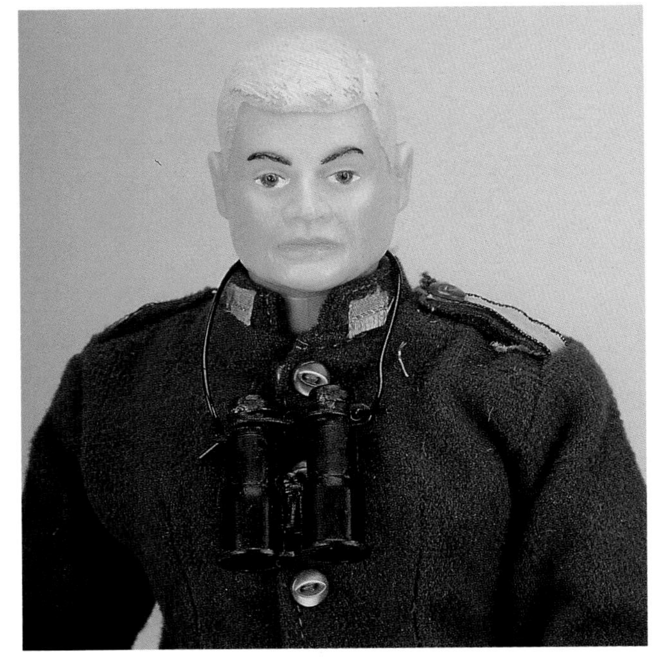

A good close-up of the sourpuss Foreign head, worn by all figures but the Japanese, and with a variety of hair colors. *Courtesy of Play With This.*

Russian, Aussie, and French Resistance. (My idea of French resistance is leaving a cheap tip at Le Cirque.) *Courtesy of WO1 Neil T. LaSala, USMC.*

Boxed "large box" Russian. *Courtesy of Roddy M. Garcia, Houston, Texas.*

Boxed "large box" Aussie. *Courtesy of Roddy M. Garcia, Houston, Texas.*

French Resistance Fighter "large box" version. Some boxed foreign figures have regular Joe heads. *Courtesy of Roddy M. Garcia, Houston, Texas.*

Japanese, German, and British ... the Legion Of Doom! (Don't think the British are bad guys? Try sitting through *Blake's Seven*!) *Courtesy of WO1 Neil T. LaSala, USMC.*

Above left: Japanese Imperial Soldier in "large" box. *Courtesy of Roddy M. Garcia, Houston, Texas.*

Above: German Soldier (originally announced as Storm Trooper) in large box. *Courtesy of Roddy M. Garcia, Houston, Texas.*

Left: British guy in "large" box. *Courtesy of Roddy M. Garcia, Houston, Texas.*

The Cycle, mint in package, made by Irwin under license from Hasbro. Looks like a win for their side! *Courtesy of Play With This.*

The cycle loose, mint, and complete.
Courtesy of WO1 Neil T. LaSala, USMC.

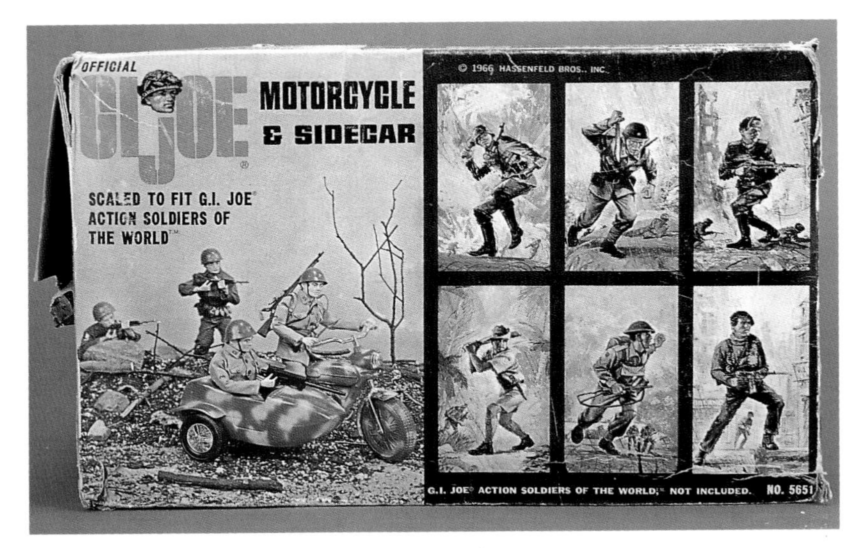

In my books you see cool box backs, too! *Courtesy of Play With This.*

The box bottom shows other Irwin vehicles. *Courtesy of Play With This.*

Certain items were issued in grab bags called Backyard Patrol bags. Here's the uniform and helmet for the Japanese Imperial Soldier. *Courtesy of Fred Mahn.*

Here's a close-up of the Japanese backpack. *Courtesy of Fred Mahn.*

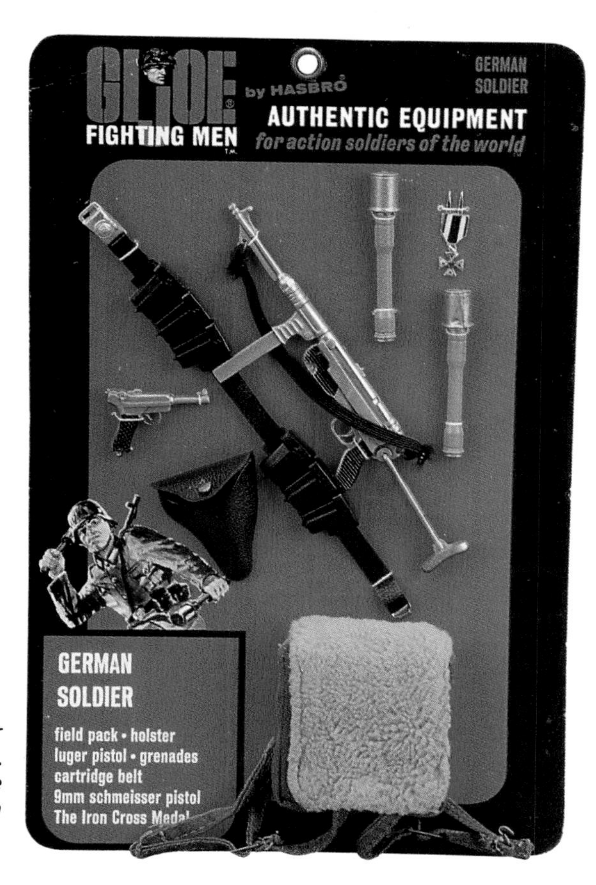

The German Soldier accessory card. *Courtesy of The Toyrareum, Ocean City, New Jersey.*

A nice detail of the German medal and "potato mashers." Militaristic types have such cute names for deadly devices. *Courtesy of The Toyrareum, Ocean City, New Jersey.*

More equipment closeups. The detail work was what separated G.I.Joe from the imitators *back then*. *Courtesy of The Toyrareum, Ocean City, New Jersey.*

Detail of the Forward Observer set, a Sears exclusive. *Copyright 1966 Sears, Roebuck, and Co.*

Now we'll have a series of shots of museum-quality examples of hard-2-find pieces. Here's the Shore Patrol sea bag. *Courtesy of Fred Mahn.*

SP

Here's a Shore Patrol arm band. *Courtesy of Fred Mahn.*

And the striped version of the Shore Patrol helmet. *Courtesy of Fred Mahn.*

Here's the West Point Cadet jacket. *Courtesy of Fred Mahn.*

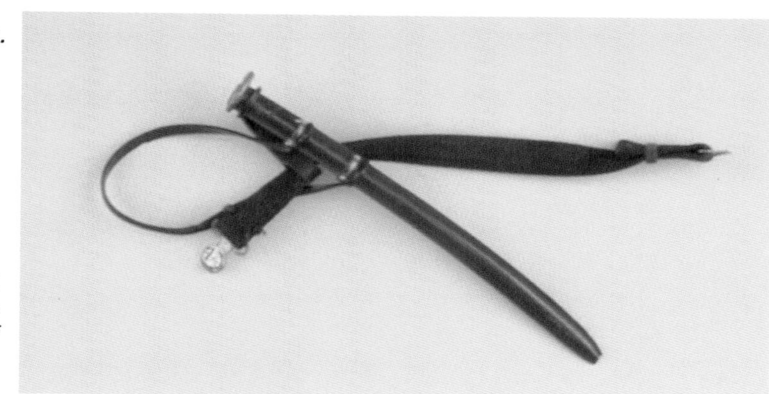

Here's the scabbard from the Annapolis Cadet. *Courtesy of Fred Mahn.*

And the Annapolis Cadet Jacket! *Courtesy of Fred Mahn.*

One of the most desired pieces is the black M.P. radio. *Courtesy of Fred Mahn.*

Conversely, some of the most common accessories are the Dress Marine jacket and pants. *Courtesy of Fred Mahn.*

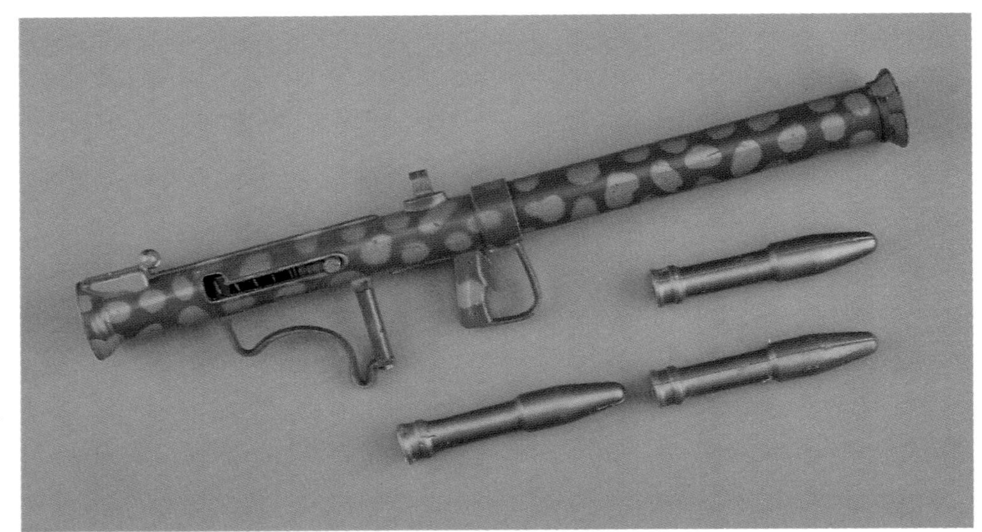

Here's the bazooka and shells. *Courtesy of Fred Mahn.*

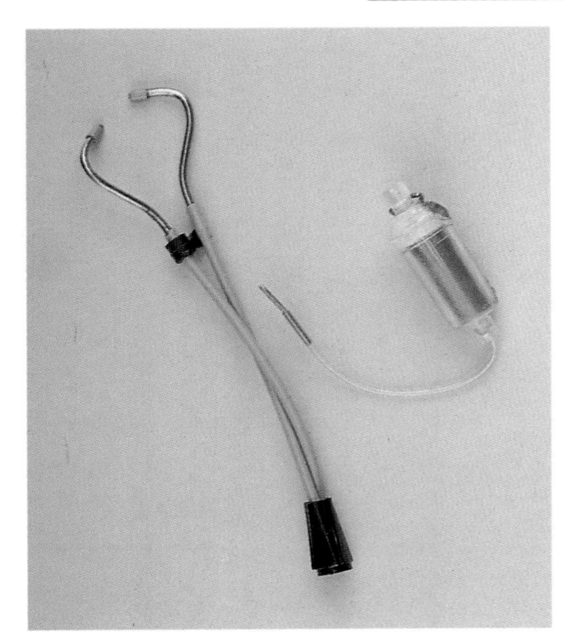

And for its victims you'll need the stethoscope and plasma bottle, complete with needle. *Courtesy of Fred Mahn.*

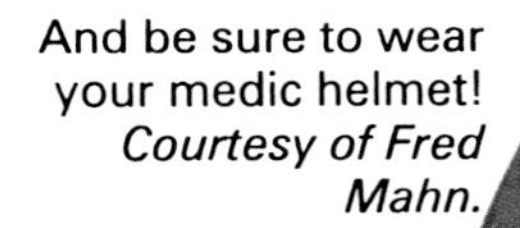

And be sure to wear your medic helmet! *Courtesy of Fred Mahn.*

The "leather" jacket with "fleece" collar from the "Tank Commander" outfit. *Courtesy of "Fred" Mahn.*

The soft plastic helmet with visor, microphone, and sticker, from the Tank Commander outfit. *Courtesy of Fred Mahn.*

The radio from the Tank Commander uniform is so rare, we thought you'd like to see it even with only one tripod leg. *Courtesy of Fred Mahn.*

From Neil LaSala's living room come all three M.P. variations: tan, green, and the far more common brown. *Courtesy of WO1 Neil T. LaSala, USMC.*

Closup of the "tan" M.P. *Courtesy of WO1 Neil T. LaSala, USMC.*

Closup of the "green" M.P. *Courtesy of WO1 Neil T. LaSala, USMC.*

The Sears Catalog section that introduced Stony to the world. *Copyright 1965 Sears, Roebuck, and Co.*

Sensational Stony section steams the sixty-six Sears sales system. Pretty good! *Copyright 1966 Sears, Roebuck and Co.*

This 1966 catalog page shows all the little Tigers and their clubhouse, a Sears exclusive headquarters with an exclusive dress uniform. *Copyright 1966 Sears, Roebuck and Co.*

Detail of all eight of Topper's Tigers. *Copyright 1966 Sears, Roebuck, and Co.*

BATTLE ACTION by IDEAL
Bridge explodes, tree falls, machine gun chatters . . in Sears exclusive Ambush Set $6.66
Get all the sound and action of full-scale combat—in one set. U.S. soldiers advance onto 36x20-inch enemy terrain. Two 15-inch 3-dimensional settings are spring activated. Jeep vehicle triggers road block—a tree topples and vibrating machine gunner opens fire. Truck in rear rumbles onto device that explodes mined bridge into 2 flying sections. Plastic. 27 men, 2 vehicles.
49 N 5970—Shipping weight 3 pounds 3 ounces. . . . . Set $6.66
BOOMM
BOOMM
RAT-TA-TAT
BANG
BANG
Soldiers pop up, fire rifles from Sniper Post $3.77
Five soldiers are mounted on heavily fortified parapet. Pull cord on spring-activated motor. They individually spring up in staggered sequence, open fire, drop for cover, then pop up again. 15x9-inch plastic setting, 8 additional soldiers.
49 N 6075—Wt. 2 lbs. 1 oz. $3.77
66c
44-piece "Fighting 69th" Set. Includes six 3-inch infantrymen plus 38 detachable combat accessories. Realistically detailed plastic, fine hand coloring.
49 N 5974—Wt. 5 oz. 66c
66c
45-pc. "Carlson's Raiders" Set. Includes six 3-inch U.S. Rangers plus 39 combat accessories that snap on and off. Realistically detailed, hand-colored plastic.
49 N 5977—Wt. 5 oz. 66c
RAT-A-TAT
RAT-A-TAT
66c
35-piece "Screaming Eagles" Set. Includes six 3-inch paratroopers plus 29 accessories that snap on. All hand-colored plastic.
49 N 5975—Wt. 5 oz. 66c
66c
31-piece "Devil Dogs" Set. Includes six 3-inch Marines plus 25 combat accessories that snap on and off. Hand-colored plastic.
49 N 5976—Wt. 5 oz. 66c
Sound and action of a Machine Gun Nest $3.68
Stranded machine-gun team waits in burned-out house ruin. At the first sign of danger, pull cord on spring-activated motor. Both let loose scathing bursts of gun fire. Machine guns actually vibrate and chatter when fired. 13½x8-in. plastic setting with 8 additional soldiers. Requires no batteries.
49 N 6074—Wt. 1 lb. 9 oz. $3.68
488 SEARS
Four Soldier Sets above made in British Hong Kong

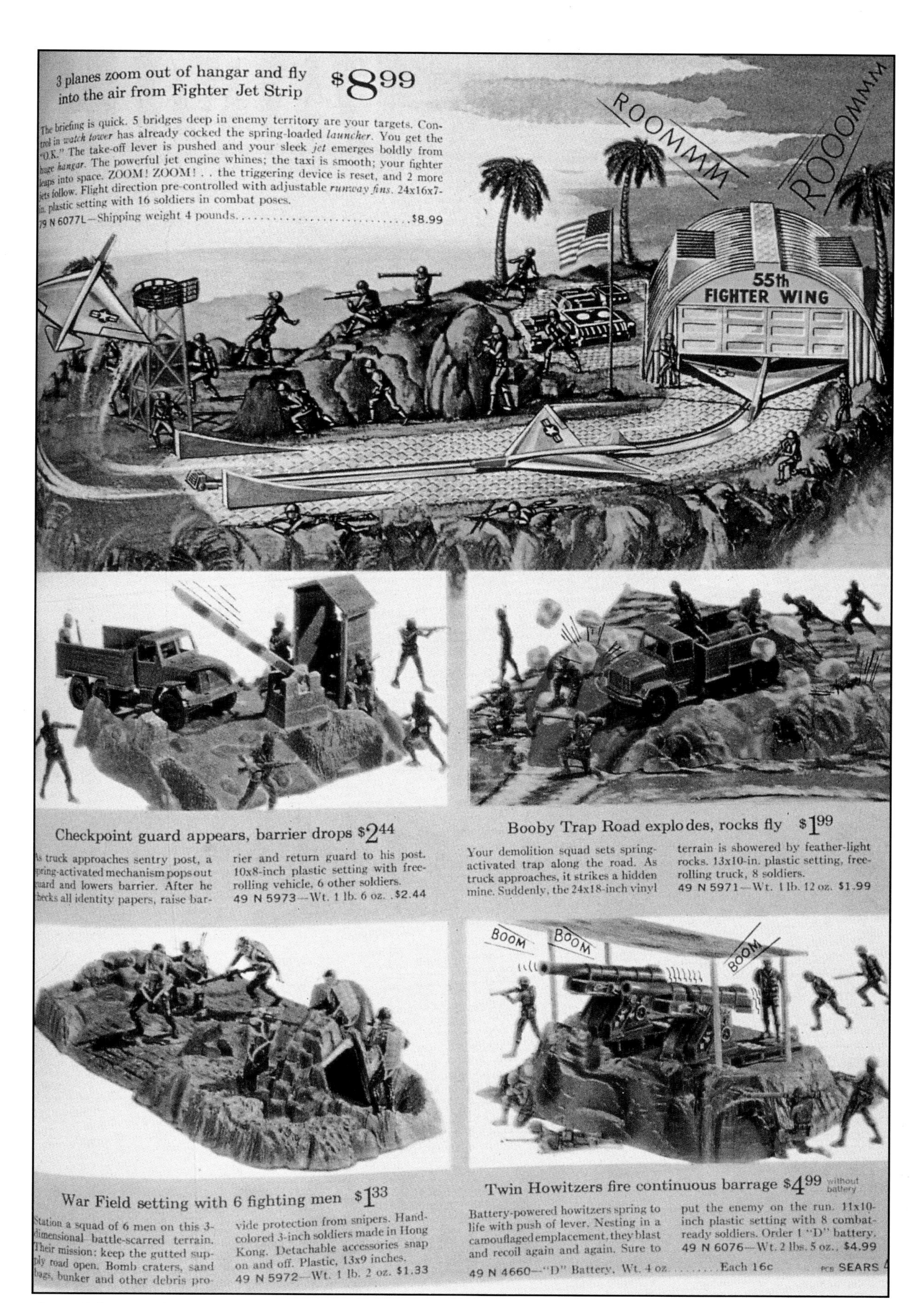
3 planes zoom out of hangar and fly into the air from Fighter Jet Strip
$8.99
The briefing is quick. 5 bridges deep in enemy territory are your targets. Control in watch tower has already cocked the spring-loaded launcher. You get the "O.K." The take-off lever is pushed and your sleek jet emerges boldly from huge hangar. The powerful jet engine whines; the taxi is smooth; your fighter leaps into space. ZOOM! ZOOM! . . the triggering device is reset, and 2 more jets follow. Flight direction pre-controlled with adjustable runway fins. 24x16x7-in. plastic setting with 16 soldiers in combat poses.
79 N 6077L—Shipping weight 4 pounds......$8.99
ROOMMM
ROOOMMMM
55th FIGHTER WING
Checkpoint guard appears, barrier drops $2.44
As truck approaches sentry post, a spring-activated mechanism pops out guard and lowers barrier. After he checks all identity papers, raise barrier and return guard to his post. 10x8-inch plastic setting with free-rolling vehicle, 6 other soldiers.
49 N 5973—Wt. 1 lb. 6 oz. .$2.44
Booby Trap Road explodes, rocks fly $1.99
Your demolition squad sets spring-activated trap along the road. As truck approaches, it strikes a hidden mine. Suddenly, the 24x18-inch vinyl terrain is showered by feather-light rocks. 13x10-in. plastic setting, free-rolling truck, 8 soldiers.
49 N 5971—Wt. 1 lb. 12 oz. $1.99
BOOM
BOOM
BOOM
War Field setting with 6 fighting men $1.33
Station a squad of 6 men on this 3-dimensional battle-scarred terrain. Their mission: keep the gutted supply road open. Bomb craters, sand bags, bunker and other debris provide protection from snipers. Hand-colored 3-inch soldiers made in Hong Kong. Detachable accessories snap on and off. Plastic, 13x9 inches.
49 N 5972—Wt. 1 lb. 2 oz. $1.33
Twin Howitzers fire continuous barrage $4.99 without battery
Battery-powered howitzers spring to life with push of lever. Nesting in a camouflaged emplacement, they blast and recoil again and again. Sure to put the enemy on the run. 11x10-inch plastic setting with 8 combat-ready soldiers. Order 1 "D" battery.
49 N 6076—Wt. 2 lbs. 5 oz., $4.99
49 N 4660—"D" Battery. Wt. 4 oz........Each 16c
SEARS

These soldiers are a good example of the kind of early, semi-action figures that qualify for this book. *Copyright 1965 Sears, Roebuck and Co.*

This detail of Ideal's War Field shows the work that went into these sets. *Copyright 1965 Sears, Roebuck and Co.*

Aside from their faces, the Tigers all had the same body. But this, too, was differentiated from G.I.Joe in that the Tigers were equipped with spring-powered right arms that allowed them to raise their guns, pick up walkie-talkies, or throw what Topper amusingly calls "pineapples."

And now, Roll Call!!!

Sarge, the leader, looked for all the world like Sterling Hayden's cigar-chomping Jack D. Ripper from Stanley Kubrick's awesome black comedy *Dr. Strangelove, Or, How I Learned To Stop Worrying And Love The Bomb*. (I am a big, big fan of that movie. In fact, I was considering titling this book Dr. Strained Pun, Or, How I Learned To Stop Groaning and Love John Marshall.) Anyway, better that the Sarge should resemble Sterling Hayden than Sterling Holloway (the voice of Winnie The Pooh).

Sarge's team was a veritable freak show of fantastic facial features. Tex was the spitting image of America's "poet lariat" Will Rogers, and had rifle-raising action. Bugle Ben was a James Dean type, and, perhaps fittingly, was equipped not with weapons—to fight *someone else's* battles—but rather a bugle that he could raise to his lips with his spring-lever arm action. Pretty Boy, meanwhile, was a Robert Stack type whose hand raised in a salute.

The Rock looked like Kenneth Tobey, Jock Mahoney, Jeff Morrow, and every other square-jawed, wood-scented leading B-man. To the accompaniment of the slogan "Try This!" written on his package, the Rock threw a "pineapple" at enemy troops.

Big Ears, the communications expert, looked kind of like a macho, balding Bing Crosby. He was able to raise a field phone to his size-enhanced ear. Combat Kid was a young spud, and looked a lot like Nick Adams, complete with squint. His arm action allowed him to throw a Molotov Cocktail and the slogan on his box actually read "Here's a Hot One For Ya!" I can actually picture good old Nick Adams saying that in a war movie. "Here's a hot one for ya, Tojo, straight outta Yancy Street!" Yoshio Tsuchiya better duck! And I thought they were friends. (Japanese sci-fi fans will love that one.)

And, finally, there was a character called Machine Gun Mike, whose face resembled sixth Stooge Joe DeRita (the one from the awful feature-length kiddie movies of the 1960s). To be specific, he resembled Joe DeRita if Joe had had a few too many brewskies and the boys had worked him over a little. Just the idea of releasing any kind of toy wearing a mug like that deserves a round of applause for audacity. Put the book down right now and clap. I'll wait.

Okay.

The Tigers also had accessories, believe it or not. Their battery-operated Tiger Tank was actually a reissue of a 1962 toy, but it looked good and was more-or-less in scale with the boys. It was activated by a remote-control box shaped like a walkie-talkie.

The Tigers Artillery Cannon and Wall Play Set consisted of—wait for it—an artillery cannon and a wall. The cannon was a massive 21" long, and it fired gray plastic shells at a breakaway plastic wall. It, too, was more-or-less in scale with The Tigers, but I'm not sure if it was a rerelease of an earlier toy or not.

So, you may be wondering, "These sound great! Why haven't I heard more about them? Why weren't they successful?"

Oh, did I forget to mention? Well, G.I.Joe was a 12" tall figure, Stony Smith was a 12" tall figure, and the Topper Tigers were SEVEN INCHES TALL!!! Perhaps they were actually scaled to accommodate the already-exist-

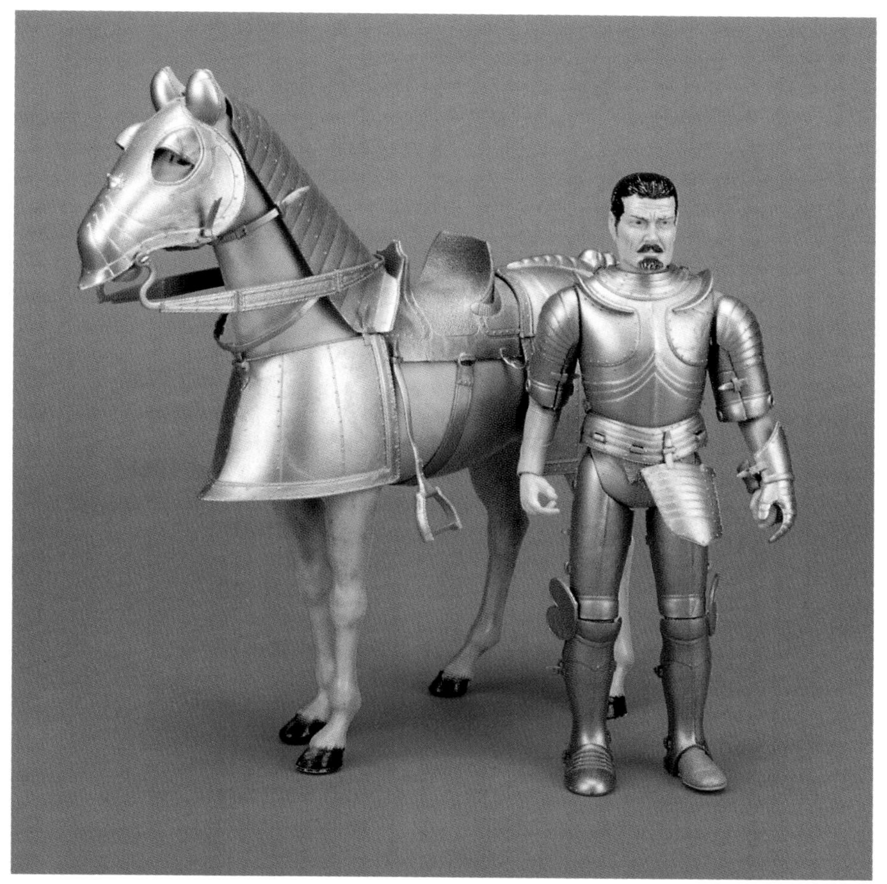

Marx's Silver Knight and his horse, with just some of their accessories. *Courtesy of The Toyrareum, Ocean City, New Jersey.*

ing tank, but, I mean, come on! Any manufacturer worth his salt knows that you have to make your figures in scale with the other toys out there or the kids won't play with 'em! "And youse can take DAT from Yancy Street, ya bums!"

And speaking of tough guys, Marx produced a highly-detailed figure of General Eisenhower as "General Of The Army." It was designed and packaged much like a G.I.Joe.

There is one more World War Two-related line that deserves honorable mention, even though technically it wasn't an action figure line. It was a series of dioramas with play soldiers and action features. It was Ideal's Battle Action series.

This series included nine different diorama scenes: Fighter Jet Strip, Check Point, Sniper Post, Twin Howitzers, Booby Trap Road, Mined Bridge, Machine Gun Nest, War Field, and Road Block, plus there was also a separate set of Fighting Men. Road Block featured a spring-activated falling tree. The tantalizingly-titled Mined Bridge had a trigger ingeniously mounted in the cobblestone road leading up to it. A rolling jeep came with over the trigger and KABLOOIE went the bridge! This was as action-filled a play adventure as any "proper" action figures could have provided.

As long as man continues to throw, shoot, swing, or spit something at his fellow man, war action toys will continue to resurface periodically in popularity. But the Sixties were perhaps the golden age of unsullied, unapologetic, military action figures!

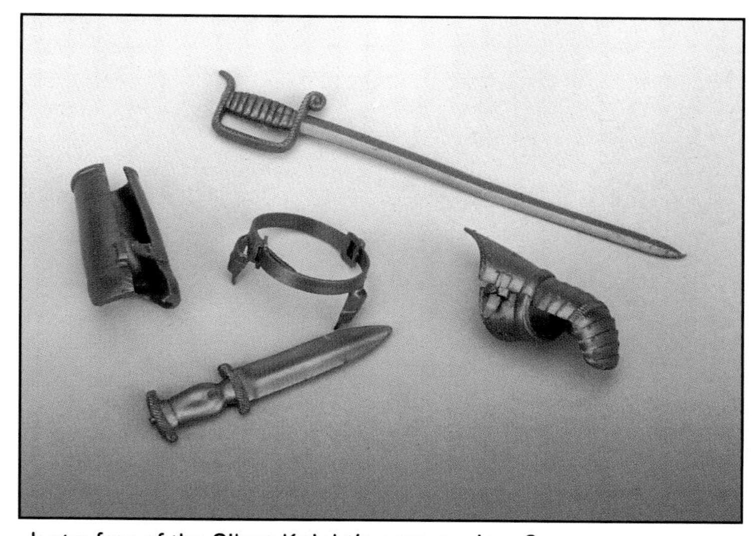

Just a few of the Silver Knight's accessories. *Courtesy of The Toyrareum, Ocean City New Jersey.*

Warhawks Price Guide

Stony Smith (Marx 1965)	LMC	MIP
Stony Smith, Original No-Knees	$95	$150
Stony Smith Paratrooper	$125	$175
Buddy Charlie, all branches	$50	$85
Stony's Jeep	$125	$175
Jeep with Searchlight	$150	$225
Carded Accessories, general range	$10-$20	$25-$35
Marx Vikings		
Blonde Sissy (Erik)	$65	$125
Brown Bearded He-Man (Odin)	$65	$125
Marx Knights		
Gold Knight (Sir Gordon)	$95	$200
Silver Knight (Sir Stuart)	$95	$200
Black Knight (Europe)	$150	$400
Gold Knight's Horse (Bravo)	$40	$100
Silver Knight's Horse (Valor)	$40	$100
Black Knight's Horse (Valiant)	$75	$200
Castle with Draw Bridge	$200	$400
Marx Rat Patrol Jeep With Figures	$900	$1500
Marx General Of The Army, Eisenhower	$150	$300

TOPPER TIGERS		
Sarge	$50	$95
Big Ears	$40	$85
The Rock	$45	$90
Tex	$40	$85
Bugle Ben	$40	$85
Pretty Boy	$45	$90
Combat Kid	$45	$90
Machine Gun Mike	$45	$90
Topper Tank	$65	$125
Artillery Cannon and Wall	$75	$150
Club House Set (Sears)	$200	$400
Battle Action (Ideal 1965)		
Ambush Set (Sears)	$50	$125
Fighter Jet Strip	$50	$95
Check Point	$50	$95
Booby Trap Road	$50	$95
War Field	$50	$95
Twin Howitzers	$50	$95
Sniper Post	$50	$95
Machine Gun Nest	$50	$95
Mined Bridge	$50	$95
Road Block	$50	$95
Ambush Set (Sears, Mined Bridge/Road Block)	$100	$200

The Silver Knight's head, *courtesy of The Toyrareum*, was the same one used for Sam Cobra, the villain in Marx's Johnny West line. And if you'll just turn the page...

Chapter Three

Wild, Wild Western Figures

"Howdy, Ma'am! Ah recognize yuh from the inner-duction," says the king of Western action figures, Marx's Johnny West. *Courtesy of Play With This.*

Hasbro had conquered the toy world with G.I.Joe in 1964. But Marx, the ages-old grand master of the toy industry, was no sleeping giant. Having produced a low-priced competitor for Joe called Stony Smith, Marx looked to the past to give the world an entirely new action figure line, one that was based on one of the oldest forms of play known in America.

Cowboys and Indians!

Marx had built its reputation on a series of playsets that featured Western themes. These, of course, were enormous dioramas consisting of toy soldier-style figures plus horses, cannons, buildings—in short, an entire world. Now that larger, poseable figures were the wave to ride, Marx looked around for Western-themed characters to license for manufacture as action figures.

Their first attempt after Stony Smith was a Daniel Boone figure, based on the Fess Parker TV series. What eventually happened, though, was that the license was dropped, and the figure was altered before being released, so as not to resemble the TV actor. This will not be the last time you'll hear me say that in this chapter.

In order to gear up for its huge new Western expansion, Marx shifted its production facilities from Erie, Pennsylvania, to more spacious quarters in Glen Dale, West Virginia. And so, in 1965, the Johnny West line was born.

The original figures were Johnny West himself (who had Stony Smith's face), Chief Cherokee, and Johnny's horse Thunderbolt. The original figures' bodies were molded in brittle, caramel-colored plastic and are hard to find today in one piece. Aside from having their clothes sculpted right on, the characters were not as well articulated as G.I.Joe. Still, they had turning heads and moveable shoulders, elbows, wrists, hips, and knees. One real advantage that Johnny had over G.I.Joe (at the time) was that his hands were cast in soft vinyl and they were sculpted so that they always looked natural, whether holding a six-shooter or a coffee pot. And speaking of coffee pots, the real genius of Marx's figure marketing was the inclusion of a large number and variety of accessories with each doll, nearly two dozen in some cases. The drawback, however, was that these accessories were molded in only one color, with no paint detailing. This was a Marx tradition handed down from their playsets. Quantity over quality was the hallmark here ... not that Johnny's accessories were exactly junk, though!

In another cash-saving move, Johnny's name came from a playset whose title Marx had already trademarked but never released—the Johnny West Ranch, because Louis Marx was never one to waste anything. The Wild West theme hit a bull's-eye for Marx, so, in 1966 the line

Jane West in an early-issue artwork box. She will kneel. It says so right on the box. *Courtesy of Sherry L. Snyder.*

Jane West in a later 1960s box, with a photo of the doll against a minimalistic art backdrop. *Courtesy of Play With This.*

was expanded to go after the girl's market. Jane West, the new addition, was a no-nonsense cowgirl who carried branding irons and a whip along with her lipstick.

The family was completed by four teenagers. The teenage cowboys, Jay and Jamie, were equipped similarly to Johnny. The girls, Janice and Josie, had Jane's accessories, minus the whip (for obvious reasons).

Thunderbolt the horse was not alone either. The old gal had managed to pop out a Thundercolt (sold separately or with his Maw). A Pinto called Storm Cloud was actually a repainted version of Thunderbolt. A new horse, Flame, was molded in a running position (through this period, the horses were solid statues). A horse named Buckskin retained the standing position, but he had a moving head. The only fully-articulated horse was Commanche, and he was part of a brand-new subset of the Johnny West line called The Fort Apache Fighters.

This offshoot series was composed of characters that had originally been licensed from TV but, for one reason or another, never made it to stores in their intended guises. There was, for instance, to be a line based on *The Wild, Wild West*. But Robert Conrad's head actually wound up on Union cavalry officer Captain Maddox in the Fort Apache series. The head of TV's Jesse James, as portrayed by Chris Jones, wound up on Sergeant Zeb Zachary. Meanwhile, Bill Buck, the cavalry scout, was the old Fess Parker Daniel Boone head on the body intended for Jesse James!

Maybe Chris Jones and Fess Parker complained, because "their" figures, Zachary and Buck, saw limited distribution. Robert Conrad was probably too busy chasing Deanna Lund or Marta Kristen around the studio lot to complain about his face being used on Captain Maddox.

Geronimo, the Apache Chief, provided a calculating adversary who pitted his wits against the Fort Apache Fighters. He was joined by the ultracool Fighting Eagle, a Mohawk (complete with punk haircut) who was apparently on vacation from his traditional stomping grounds in New York! Oh well, it was worth it to have him.

When ABC ran Fox's *Legend Of Custer* TV series, Marx was ready to give the world a George Custer in the image of Wayne Maunder, star of the show. But the series bombed (it was 1967 and the show was too war-oriented), so the Fox copyright was removed from the Marx figure. Custer was also removed—to Arizona as part of Marx's Fort Apache Fighters. At least he was left intact and suffered not the indignity of body-swapping like the other guys!

There was also an actual Fort Apache playset, which, while a lot of fun, was reportedly not as in scale with the figures as the other playsets in the Johnny West line. Oh, you know, the playsets I haven't mentioned until now! One was the Circle X Ranch, the other was the Bunkhouse, both of which could be opened out into floor-

A full page of Western wackiness. *Copyright 1966 Sears, Roebuck, and Co.*

Janice West? My mother's name is Janice! *Courtesy of Play With This.*

covering, expansive dioramas. I remember my cousin Edwin had a Circle X Ranch (probably handed down from his older brother Glenn) and I just thought it was the coolest thing in the world.

Plus, to guard the family, Marx produced the world's first dog action figures. Flick was an English Setter and Flack was a German Shepherd. I wish I had discovered Flick sooner, because our beloved family pet Squire was an English Setter, and it would have been great to add Squire to my G.I.Joe team.

Another survivor from TV who was spared the Fort Apache treatment was none other than lovable lush Dean Martin, as Sheriff Pat Garrett. He got to stay in character, although his body was just a recycled Jesse James body with a sheriff's star sculpted on to it. There's no truth to the rumor that Pat Garrett's accessories included a martini shaker and a little black book with Joey Heatherton's phone number in it.

And speaking of feminine female women, the Indians got one of their own in the form of Princess Wild Flower. She was one of the better-detailed and most realistic figures of the line. The Princess apparently was not well-guarded, for she carried a papoose along with her other accessories. For a Marx figure, she featured an impressive variety of colors in her accessories. Other "Indiana" made for the warriors of the west included a vinyl teepee to scale with the figures and an Indian canoe—a vacu-formed, two-foot-long, orange plastic stealth machine. When not battling Caucasoid invaders, Geronimo, Chief Cherokee, and Fighting Eagle could cruise down the river in search of the elusive buffalo figure that Marx saw fit to create for the line.

Plus, at long last, Johnny West was given a real bad guy to fight. Sam Cobra came dressed in black (well, okay, molded in black) and his accessories included at "hat-hide-able" derringer, skeleton keys, and even a pool cue, so you knew he was bad!

Marx also experimented with smaller dolls for a short period of time. These were a line of 7-8" Western series figures (scaled with their Rat Patrol figures and NASA Astronaut) including a cowboy named Johnny Colt and a Geronimo-like Indian called Red Cloud.

The last of the good guys to be produced was a member of the oft-neglected African-American settlers: Jed Gibson, a cavalry scout. He is among the rarest figures in the series and is probably the most sought-after.

Basically, from 1967 through the end of the decade, the Johnny West line continued on with no major additions or changes. The boxes, though, were constantly changing. Early on, they were illustrated with artwork representing the characters. Later, the figures themselves were shown in photos, augmented by artwork backgrounds that changed stylistically over the years. Eventually, changes were made to the line itself. Some of the figures were dropped, and Johnny and Sam Cobra were given Quick Draw Action.

Josie West? My mother's name is—oh, wait, no it isn't. *Courtesy of Play With This.*

Unlike any other action figure introduced in the 1960s, Johnny West rode high, virtually unchanged, well into the 1970s. And with great sales, to boot.

Cowboy boot, that is.

Alongside Johnny, there was another set of western figures of similar construction to the Marx line. These were the Bonanza figures from American Character. Although they stood a few inches shorter than Johnny, they were just as tall in the saddle. The assortment included Ben Cartwright, Hoss, and Little Joe. Originally, there was to be a Parnell Roberts figure, but the actor left the series. So American Character slapped an evil mustache on him and released him as The Outlaw—predating Sam Cobra by a year or two.

The horses, though crafted as solid statues, had ball bearings on their feet so they could roll along the prairie. You could buy them separately. Ben, Hoss, and Little Joe also came in deluxe boxes with their horses. There was also a four-in-one prairie wagon that came with over five dozen pieces.

The Bonanza boys were so popular, in fact, that Revell released a model kit featuring the three standing figures of Ben, Hoss, and Little Joe. Both types of Bonanza figures sold out quickly. Thus, many harried store employees were forced to concede, "Yes, we have no Bonanza, we have no Bonanza figures today."

Although Westerns never were as popular again in the 20th century as they were up through the 1960s, that hasn't diminished the collector appeal of these atmospheric and well-crafted figures.

Jamie West, The Adventures Of Johnny West When He Was A Boy! No, not really. *Courtesy of Play With This.*

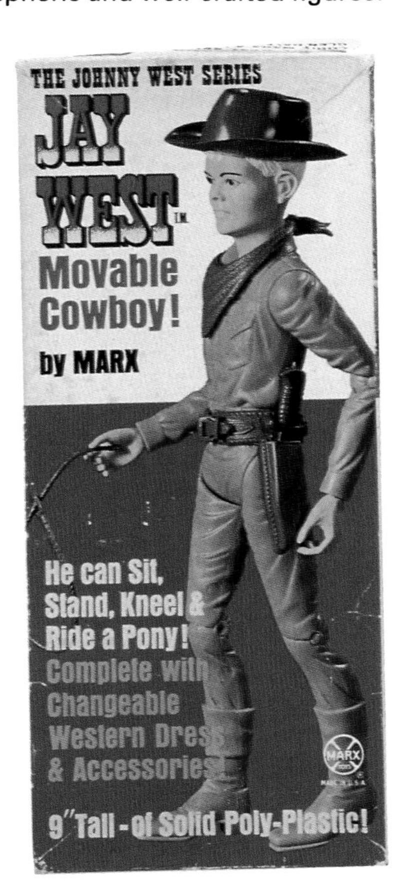

Jay West. My parents called me Jay when I was little and then Mark Lumadue called me that as an adult after Jay Gatsby, a character in a book that is not about toys. *Courtesy of Play With This.*

Western Action Figures Price Guide

Johnny West/ Best Of The West	LMC	MIP
Johnny West, first issue box	$45	$125
Johnny West, later boxes	$45	$95
Johnny West, Quick-Draw Action	$40	$75
Jane West, first issue box	$25	$75
Jane West, later boxes	$25	$60
Janice West	$25	$50
Josie West	$25	$50
Jay West	$25	$50
Jaimie West	$25	$50
Geronimo, first issue box	$35	$95
Geronimo, later boxes	$35	$65
Chief Cherokee	$35	$65
Fighting Eagle	$55	$95
Princess Wild Flower	$45	$95
Sam Cobra	$40	$85
Sam Cobra w/ Quick-Draw Action	$45	$95
Sheriff Garrett	$40	$90
General Custer	$40	$90
Captain Maddox	$40	$90
Zeb Zachary	$45	$95
Bill Buck	$45	$95
Joe Gibson	$95	$225
Daniel Boone	$95	$225
Thunderbolt	$25	$60
Storm Cloud	$30	$70
Flame	$30	$60
Buckskin	$30	$60
Commanche	$45	$75
Pancho	$20	$40
Thundercolt	$20	$40
Thunderbolt w/Colt	$50	$75
Buffalo	$40	$85
Flick the German Shepherd	$30	$65
Flack the English Setter	$30	$65
Figure & Horse Combination Sets	$50	$115
Indian Teepee	$25	$50
Indian Canoe	$30	$95
Carry Cases	$25	
Circle X Ranch	$50	$95
Bunkhouse	$40	$75
Buckboard w/ Thunderbolt	$50	$95
Covered Wagon w/ Horse	$50	$95
Jeep Set	$50	$95
Bonanza (American Character 1966)		
Ben Cartwright	$45	$125
Ben's Palomino	$30	$75
Hoss	$45	$125
Hoss' Stallion	$45	$75
Little Joe	$50	$135
Little Joe's Pinto	$30	$75
Outlaw	$55	$150
Outlaw's Mustang	$30	$75
Combination Sets	$75	$200
4-in-1 Wagon	$45	$95
Revell		
Revell 3-figure Model Kit	$60	$125

Here's a highlight of the page you saw earlier, mainly to take up space. Still, you get to see the rare Corral up close! *Copyright 1966 Sears, Roebuck and Co.*

Do you ever wonder what Geronimo might have shouted if he ever jumped out of a plane? I lose a lot of sleep over things like that. Did I mention that he is strong and has many weapons? End of message. *Courtesy of The Toyrareum, Ocean City, New Jersey.*

Marx's Jeep Cherokee ready for—what? Oh, I'm sorry. Chief Cherokee. *Courtesy of The Toyrareum, Ocean City, New Jersey.*

And speaking of jeeps, here's the Sears exclusive. I thought the Johnny West line was a period piece. And look at Thunderbolt with his wife and son. *Copyright 1966 Sears, Roebuck and Co.*

capsule shaped like an inverted cone that traveled under its own power. This battery-powered vehicle could be "programmed" by placing pegs in various positions on a control panel. It came with cardboard standups of planets and stuff so you could pretend that the Star Seeker was actually traveling around the solar system.

The Star Seeker was also part of an enormous play set called The Voyage To Galaxy Three. Along with a Major Matt figure, it included a flying disc spaceship piloted by a character called Or from Orion. Mr. Or and his ship, the Orbitor, also came separately, complete with a kid-powered launcher. Mr. Or is considered to be the rarest of all Major Matt items, since he was just a little plastic blop and his Orbitor could easily be used like a Frisbee.

And speaking of aliens, as Major Matt headed into the 1970s, the emerging decade brought a new alien: Scorpio. This pink and purple, snail-headed alien had body armor and light-up eyes. He came with a chest plate that fired small "search globes" made of foam. Unlike the astronauts and Callisto, who were six inches tall, Scorpio towered over them at EIGHT INCHES. The Scorpster was not mixed into the regular figure assortments that were shipped out. Stores had to order a case of all Scorpios, a risky move for a nonhuman figure. As a result, not many places did order it, so Scorpio was, and is, hard to find.

Mattel also decided to put out a Talking Major Matt Mason. His backpack was equipped with a voice-box, pull-string, and five phrases. But instead of having a kid pull the string from Matt's body and have it retract as he spoke, a la G.I.Joe and Barbie, Matt Mason retrieved his cord while "flying" up it! The Voice Command Pack (or VCP) was removeable and could be used with other figures. The VCP even had legs on it, to help balance whichever figure was wearing it.

Another talking version of Major Matt Mason was the so-called Talking, Flying Major Matt Mason. This figure came with the awesome XRG-1 Reentry Glider—a vehicle which resembled a stealth plane. The glider had a cockpit in which to place Major Matt (without his VCP) and return him safely to Earth. The glider could only fly correctly when balanced by the specific weight of a Matt Mason series astronaut. Pretty tricky, huh?

Above: Doug Davis, the dastardly doer of detrimental deeds, came packaged on a Cat Trac. He deeply resented his yellow uniform and plotted to sabotage the missions. Fortunately, no one ever found out. *Courtesy of Play With This.*

Left: Scorpio, not just an astrological sign, but an astro-zombie as well! *Courtesy of Play With This.*

Major Matt on his cat trac. Jump back! *Courtesy of Play With This.*

Here's Scorpio in the box. I dig that freaky artwork, man. *Courtesy of Play With This.*

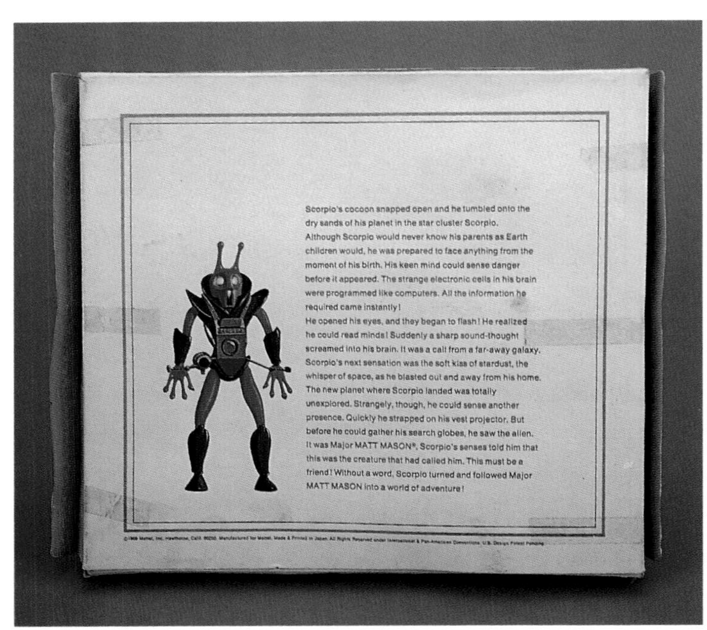

The text implies that Scorpio is a good guy, but this is the Sixties, and guys with antennae that big can only be bad! *Courtesy of Play With This.*

Reconojet Pak, wack, wack, wack! *Courtesy of Play With This.*

Probably the rarest of the late-model accessories was the Uni-Tred, a tractor-like vehicle that hauled a Space Bubble. The tinted bubble rotated 360 degrees and a figure could be seated inside. The figure inside remained perfectly upright in his Gyro-Seat. There is no discernible reason why the astronauts would want or need to ride in such a contraption, so I guess they just did it for fun, at the taxpayers' expense. Something called a Dual Action Tow Yoke held the two rigs together.

Matt Mason was a spacey selling machine and inspired all sorts of merchandise from board games to Big Little Books. But it also inspired Colorforms, who issued a line of seven "me too" bendy characters in the late 1960s, specifically to be used with Major Matt Mason. These are known as the Outer Space Men.

Alpha Seven was your proverbial Little Green Man From Mars, and he stood about three inches tall. He originally came equipped with a ray gun and a clear bubble helmet. Electron Plus, The Man From Pluto, was a robotic character reminiscent of the dude in the movie *The Man From Planet X*. He came, originally, with a bubble helmet and ray gun.

Commander Comet, The Man From Venus, was the only humanoid in the group. I guess Colorforms wanted a "good guy." He had a bubble helmet, crossbow (ooh, so high tech!) and strangely lovely angel wings on his back. Xodiac, The Man From Saturn, could have stepped right out of the Matt Mason line, as he was a fairly straightforward spaceman from the shoulders down. Ahh, but atop those shoulders sat a demonic, fiery-orange head. He also came with a bubble helmet and a weapon.

But the best three figures were the truly monstrous, evil-exuding aliens. Orbitron, The Man From Uranus, was a red monstrosity akin to the Metaluna Mutant from *This Island Earth*. He came with a weapon and an enormous, exposed brain! Astro-Nautilus, The Man From Neptune, was exactly that—an octopoid head and tentacles atop human-esque legs, armed with a Neptunian trident. But of all the Outer Space Men, the most fondly-remembered, and most ferocious, was Colossus Rex, The Man From Jupiter. He was a scaly, green, barrel-chested, bat-eared giant! His weapon was a low-tech, but highly-effective medieval style mace!

There was a second series of aliens, but they never made it past the early production stage. These included a devil-like character in a heat-insulating suit, a cyclops beast, and a strange, two-headed alien. If you're desperate to have these, well, there are a couple prototype sets floating around here on Earth, some loose and some carded. They occasionally turn up, but Buddy, they ain't cheap! These additional six characters were intended for debut at the 1969 Toy Fair. A dock strike held up the prototypes in transit and they were not able to be shown. By the 1970 toy fair, interest had waned.

And there you have a pretty good summary of all the bendy space characters of the Sixties.

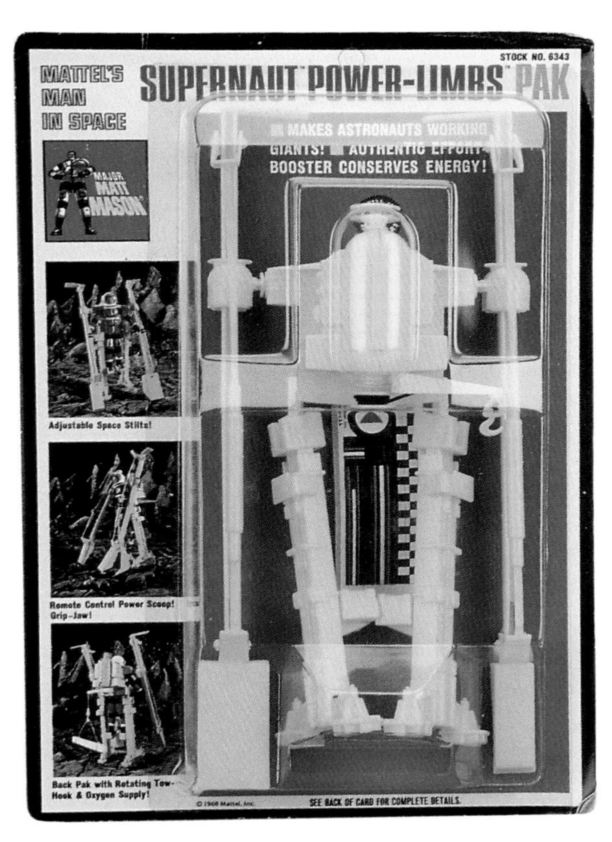

Supernaut Power Limbs, predating the yellow exo-suit in *Aliens* by two decades. *Courtesy of Play With This.*

Space Probe Pak. I love that. "Pak." There is no letter "C" in space! *Courtesy of Play With This.*

Gamma Ray Gard Pak! No letter "U" either. It's space language, I think. Sort of universal. *Courtesy of Play With This.*

Space Power Suit Pak. Yeah. *Courtesy of Play With This.*

Major Matt Mason

	LMC	MIP
Major Matt Mason	$60	$195
Talking Major Matt Mason	$95	$295
Sgt. Storm	$70	$295
Doug Davis	$70	$295
Jeff Long	$125	$495
Callisto	$95	$250
Scorpio	$350	$1500
Captain Laser	$75	$295
Mission Team 4-Pack	$350	$750
Reconojet Pak	$25	$50
Moon Suit Pak	$25	$50
Space Shelter Pak	$25	$50
Space Probe Pak	$25	$65
Satellite Launch Pak	$25	$65
Rocket Launch Pak	$35	$65
Supernaut Power Limbs	$35	$125
Gamma Ray Gard	$30	$125
Space Power Suit	$35	$125
Space Shelter	$35	$125
Space Crawler	$35	$125
Uni-Tred w/ space bubble	$95	$175
Star Seeker	$75	$195
Firebolt Cannon	$45	$95
Astro-Trac	$45	$95
XRG-1 Reentry Glider	$85	$195
Orbitor with Or Playset	$450	$1900
Satellite Locker	$45	$195
Deluxe Space Mission Set w/Matt	$195	$350

Space Travel Pak, for your toothbrush and stuff. *Courtesy of Play With This.*

Outer Space Men

Alpha Seven, The Man From Mars	$85	$350
Electron +, The Man From Pluto	$85	$350
Commander Comet, The Man From Venus	$85	$450
Xodiac, The Man From Saturn	$95	$450
Orbitron, The Man From Uranus	$95	$450
Astro-Nautilus, The Man From Neptune	$175	$700
Colossus Rex, The Man From Jupiter	$175	$700

Uni-Tred, camel of the moon. *Courtesy of The Toyrareum, Ocean City, New Jersey.*

The Germans call it a Weltraum, but we know better. Major Matt was very big in Europe. *Courtesy of Play With This.*

Series Two World of the Future (not released)
Cyclops, The Giant From
Beyond The Milky Way.
Gamma X, The Man From
The Fourth Dimension. (Glows in the dark)
Gemini, The Man From
Twin Star Algol. (Two heads)
Inferno, The Flame Man of
Mercury. (Lights up head)
Mysteron, The Man From
Hollow Earth.
Metamorpho, The Man From
Alpha Centauri. (Three interchangeable faces)

Here's the deluxe version! *Courtesy of Play With This.*

The Satellite Locker, with an "airlock" storage compartment for your favorite Astro-Nut, preferably Don Knotts or Tim Conway. *The Astro-Nuts* (Disney, 1978) is my favorite nonexistent movie. *Courtesy of Play With This.*

"I'm sending my Star Seeker to find Patty Manterola!" says every Mexican Matt Mason fan. *Courtesy of Play With This.*

Firebolt? Give me a jalapeño pizza and I'll show you a firebolt! *Courtesy of Play With This.*

Space Crawlers! They're all over me! Get 'em away! AAAIIIEEE!!! *Courtesy of Play With This.*

The Space Station allowed kids to act out cool adventures. Callisto invades the base. Captain Laser trips and knocks it over. Think of the space-possibilities! *Courtesy of Play With This.*

My favorite thing ever published in *The Fortean Times*, the magazine covering bizarre phenomena (ba tee ba teepee) was when they ran a reader survey that asked, "If there really are little men from Mars, are they really green?" Alpha Seven here thinks so. *Courtesy of The Toyrareum, Ocean City, New Jersey.*

success of the *Batman* TV show. It was just fortunate that God made super heroes popular just in time for Ideal to get a huge boost in sales.

Of course, quality had something to do with it. Cap-a-riffic's clothing and accessories were made just as well as G.I.Joe's. Cap himself, while not quite as poseable as Joe, could still be positioned realistically. Plus he had more naturally-crafted hands, cast in softer plastic that Joe's. Cap could hold things much more naturally than Joe (an advantage that Johnny West also had).

Even better, Captain Action lived in a world of ray guns, alien devices, leotards, and capes ... things that the kids who played with G.I.Joe were very much into. I mean, Filmation was producing *Superman* and *Justice League* cartoons, Hanna-Barbera had *The Fantastic Four*, and then there were those moving cut outs in the *Marvel Heroes* cartoons. Once the Filmation *Batman* cartoon arrived along with the *Spider-Man* cartoon, there were long underwear characters everywhere!

It was on this wave that Ideal rose, not only with Captain Action but with its toy soldier-style super hero figurines and playsets (see Super Freak-Out Grab Bag). Kids of the Sixties were leotard-happy and income tax sappy. They had no financial responsibilities other than to spend their money on the things they loved the most. Ideal was happy to hear it.

In 1967, Ideal introduced Action Boy, a companion figure who could become three different heroes, plus the cool Silver Streak amphibian car. There was even a DC Comics Captain Action title, wherein we learned that his real name was Clive Arno. (C.A., get it? Get it?)

Captain Action in his photo box, stepping on Dr. Evil. *Courtesy of The Toyrareum, Ocean City, New Jersey.*

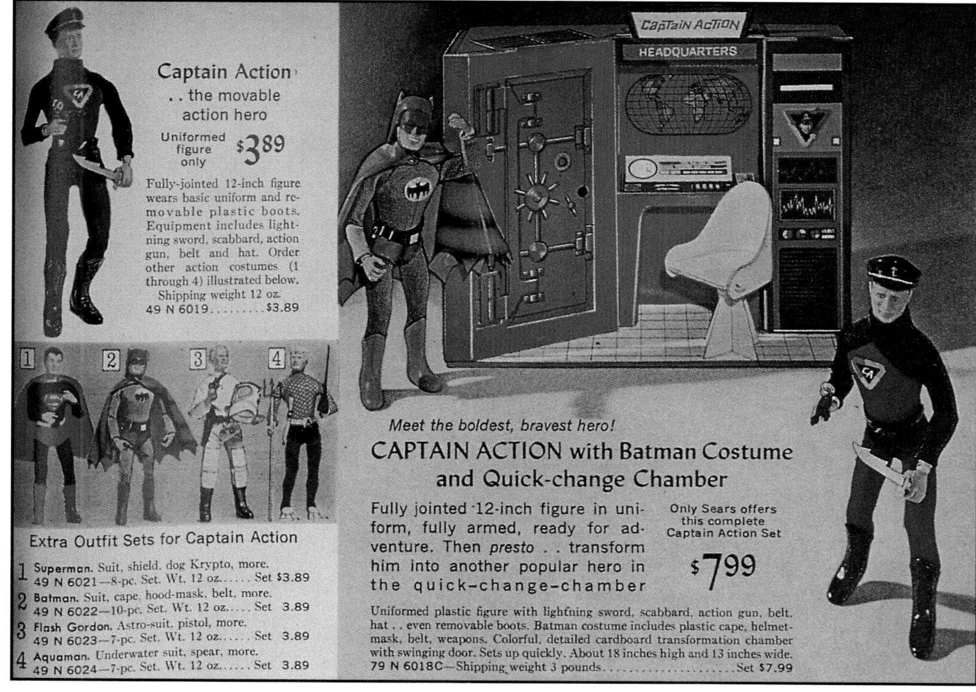

The Captain Action section of the 1966 Sears catalog, showing assorted costumes and the exclusive Quick Change Chamber combo set. *Copyright 1966 Sears, Roebuck and Co.*

The next big release was a set of four dolls based on DC Comics heroines. Ideal sold them individually as whole dolls, not as outfits sets. That is kind of a shame. If there had been an "Action Girl" it would have given girls a new hero (besides Barbie). And it would have given all of us toy collector sickos the chance to handle a busty female figure in a form fitting, blue-and-black leotard.

Oh well. Anyway, the four feminine figures were made with the same attention to detail as the Captain Action line. They also came with reused Captain Action accessories. Supergirl had Krypto and a Linda Danvers outfit. Batgirl had a Batarang and Barbara Gordon outfit. Mera (Mrs. Aquaman) had no extra outfit but came with Aquaman's trident, and box art that showed her holding her son Arthur Junior (a.k.a. Aqua Baby). The last, and best, was Wonder Woman (no, not Patty Manterola, another wonder woman) who came with Captain America's shield.

In 1968, the greatest addition of all arrived in the form of Dr. Evil, who had the dubious distinction of having his name stolen for the villain in the movie *Austin Powers, International Blah, Blah, Blah.* Dr. Evil's arrival back in 1967 was announced in the Ideal retailer's catalog with the heading: "A dramatic new chapter in the lives of Captain Action and Action Boy!" He came, as the comic book ads described, with "An Evil Outfit and Evil, Evil Things."

Each of the three dolls was issued more than once. Captain Action's original box art, showing the characters he turned into, was altered when the Lone Ranger's shirt was changed from red to blue. A third version box promised Cap had an enclosed working parachute. A fourth issue box featured a photo of Captain Action stepping on Dr. Evil's head. Action Boy's first box was similar to Captain Action's. He was reissued in a photo box, wearing a brand-new space suit. Dr. Evil's first box was part of the photo series. He was reissued in a display box with lab accessories, probably the single most sought-after Captain Action item.

Certain changes were made during the life of the line. The Sgt. Fury outfit was dropped from the assortment when Tonto, Buck Rogers, Spider-Man, and Green Hornet were added, to make an even dozen. Later, all outfits were issued with flicker rings that showed Captain Action from one angle and the costume's particular hero from the other. Since Sgt. Fury was dropped earlier, there is no flicker ring for him.

Collectors love to speculate about potential, unreleased Captain Action and Action Boy outfits like The Flash, Green Lantern, Kato, and others. Some artistic aficionados have actually created their own versions of these outfits. It would have been great to see Joker, Luthor, or Green Goblin dress-up outfits for Dr. Evil. And we would have loved Comic Villainesses like Catwoman or Poison Ivy, the original pistil-packin' mama.

One of the great mystery areas for collectors is the realm of licensed Captain Action products. There's never been, to my knowledge, a thorough listing of them. The best-known item is Aurora's model kit of him, which didn't exactly set the world on fire. After all, kids wanted the Cap-O-tron mainly to change him into somebody else. Of course, the outfits fit on G.I.Joe, too. In fact, many of the prototypes in the Ideal sales reps catalogs clearly show "Cap" sporting G.I.Joe hands!

The one licensed item that I know exists is a child-sized raft. My buddy Fred Mahn had it in his basement for years, protecting the floor from the cat box. When he got into toys, he realized what he had, hosed it off, and traded it to a toy dealer for a nice bunch of G.I.Joes and accessories.

Captain Action enjoyed three healthy years of sales. At the close of the sixties, kids were more and more fascinated by the space race. Declining interest in super heroes, plus what must have been a hefty amount in royalties to retain rights to the characters, convinced Ideal not to proceed with the Capster into the 1970s.

Captain Action did show, though, that there was a BIG market for well-made figures based on super heroes. There would be many more to come in the next three decades.

Cap dressed in his Superman outfit. *Courtesy of The Toyrareum, Ocean City, New Jersey.*

The insert card from the boxed Superman outfit, with all accessories intact. *Courtesy of The Toyrareum, Ocean City, New Jersey.*

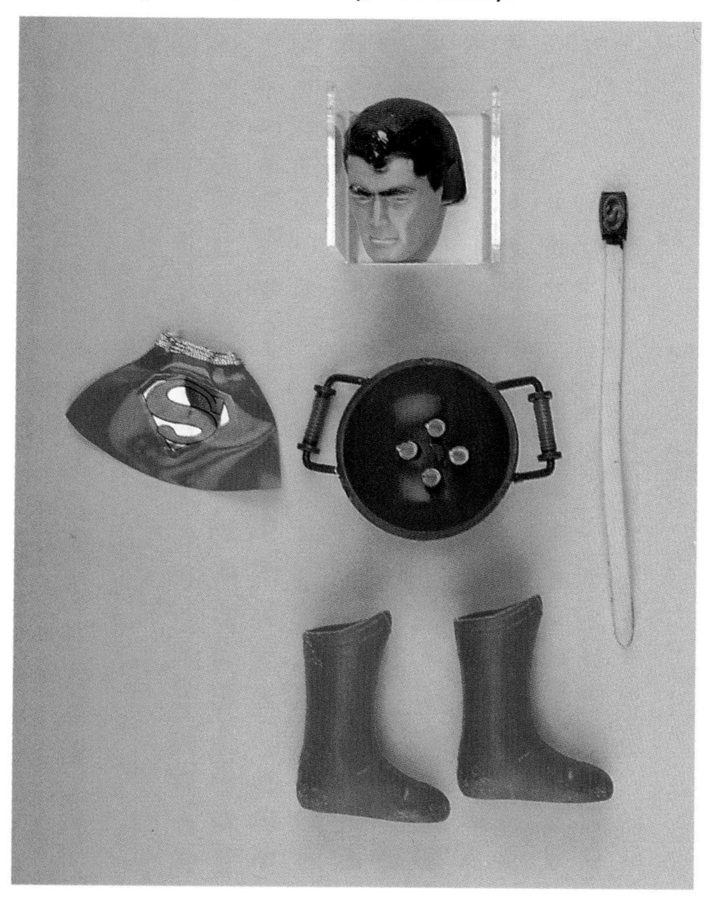

Closeups of the mask, boots, and belt, plus the Phantom Zone Projector and Krypto's cape. *Courtesy of Play With This.*

Captain Action Price Guide

Action Figures	**LMC**	**MIP**
Captain Action		
First Box (red Ranger shirt)	$125	$300
Second Box (blue Ranger shirt)	$125	$450
Captain Action, W/ Parachute In Box	$225	$650
Captain Action In Photo Box	$125	$650
Captain Action In Pieces		
Figure, nude	$95	
Hat	$10	
Leotard	$15	
Boots, pair	$10	
Belt	$15	
Laser gun	$15	
Lightning Sword	$25	
Parachute w/harness	$50	
Action Boy		
Action Boy, First Issue Box	$300	$850
Action Boy Spacesuit/ Photo Box	$500	$1100
Action Boy Deconstructed		
Figure, Nude	$125	
Leotard	$50	
Kicky Beret	$45	
Boots, pair	$30	
Belt	$20	
Boomerang	$45	
Knife	$30	
Khem the Panther	$25	
Khem's leash	$25	
Space Suit	$75	
Space Helmet	$50	
Space Boots, pair	$30	
Utility Belt	$25	
Ray Gun	$40	

Reflecting our national obsession with the Phantom Zone Projector, here is a closeup of it. *Courtesy of Play With This.*

Dr. Evil, Intergalactic Man Of Mystery		
Dr. Evil, Complete, in Photo Box	$450	$1400
Dr. Evil, Complete w/ Lab Set	$1000	$3000
Dr. Evil, His Evil Outfit, and Evil, Evil Things		
Figure desnudo	$125	
Shirt	$35	
Pants	$35	
Sandals, pair	$100	
Medallion	$150	
Laser Gun	$40	
John Marshall-looking mask	$35	
Lab Set Pieces		
Lab Set Hypno-helmet	$65	
Lab Set "Oriental" Mask	$75	
Lab Set Hypno-Eye	$60	
Lid for Hypno-Eye	$40	
Machine/base for Hypno-Eye	$75	
Ionized Hypo(dermic) Rifle	$80	
Reducer Wand with prism	$100	
Outfit Sets		
Batman		
Batman in first issue box	$150	$500
Batman with flicker ring	$175	$700
Bat-Paraphernalia		
Leotard	$25	
Cape	$40	
Mask, Face section	$15	
Mask, Hood section	$15	
Boots, pair	$16	
Utility Belt	$20	
Flashlight	$25	
Laser Beam drill	$25	
Grappling hook and retracting box	$35	
Batarang	$25	

A highlight from Sears, with a good view of the Cap in his regular outfit as well as Cap in a Batman suit. Only one doll actually came with the combo set. *Copyright 1966 Sears, Roebuck and Co.*

The Batman suit on the insert card from the box it came in. This is the later version with the flicker ring. *Courtesy of The Toyrareum, Ocean City, New Jersey.*

Cap dressed as Batman. Ironically, this is the eleventh photo in the chapter, and was the first shot on the eleventh roll of film we shot for this book. Holy Coincidence! *Courtesy of The Toyrareum, Ocean City, New Jersey.*

The Futility Belt, which for some reason is blue and not yellow. *Courtesy of Play With This.*

The Bat Drill. *Courtesy of Play With This.*

An illuminating shot of the Bat-Flashlight. *Courtesy of Play With This.*

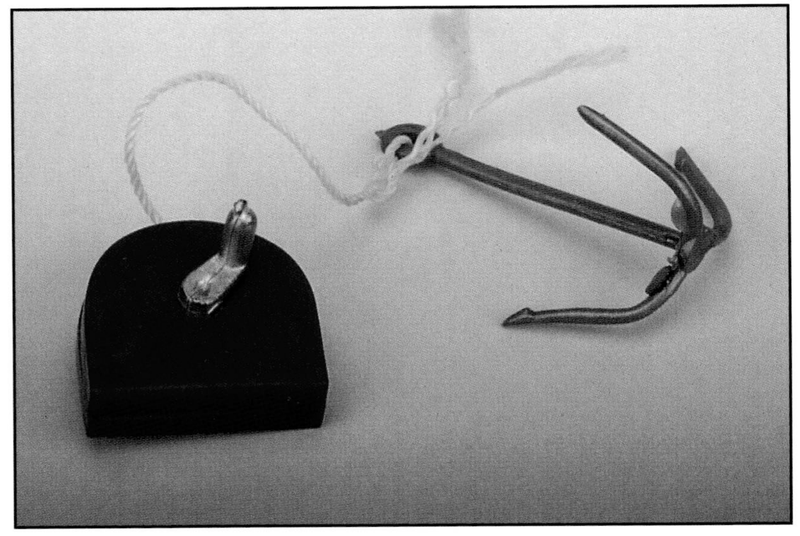

The Bat Grappling Hook, Bat Line and Bat Winder. *Courtesy of Play With This.*

I always liked the inclusion of a Batarang with this outfit, because when I was a kid in the 1970s we had Mego Batman dolls who had no accessories. Of course, the little Mego Comic Action Batman had a batarang and line, but that was a solid piece of plastic in the shape of a rolled-up rope with a Batarang on the end. Not a real Batarang like this one. Ironically, Mego considered, but never issued, a Batman playset that included a Batarang. Eh, I just like saying "Batarang." *Courtesy of Play With This.*

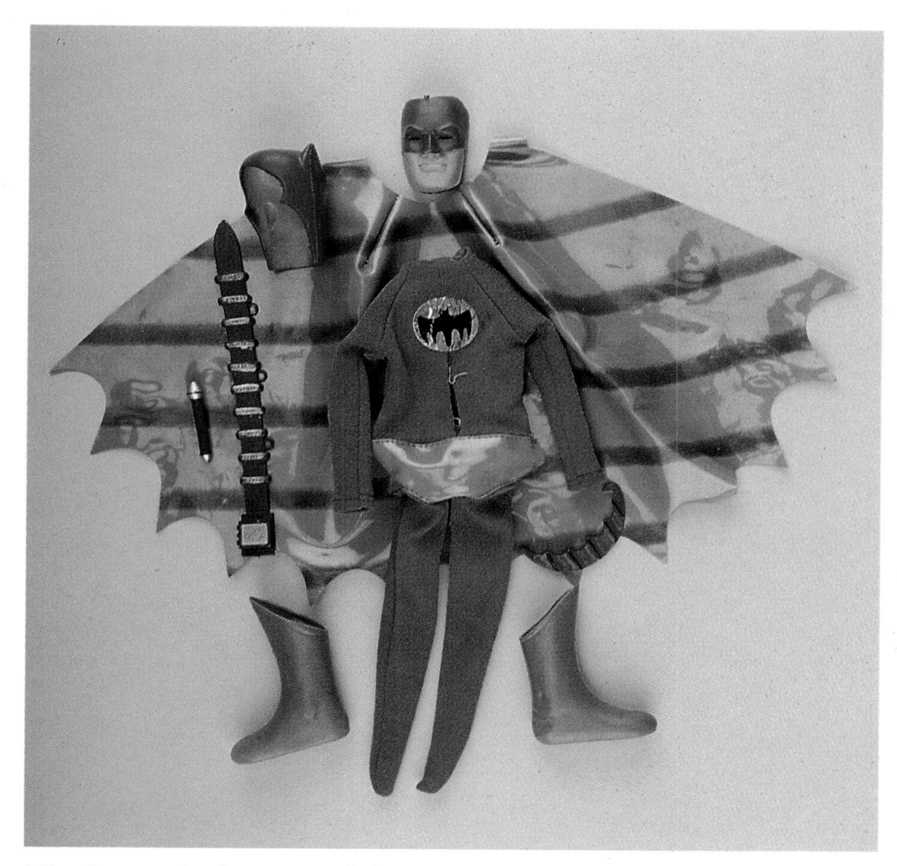

I like the way the Batman outfit looks laid out. I don't know how long it's been dead, but I like the way it's laid out. Rim shot! Thank you, I'm here all week! *Courtesy of Play With This.*

Another Sears highlight, this one showing Cap and some of his other early outfits, like Aquaman and Flash Gordon, that we have no other shots of. Narf! *Copyright 1966 Sears, Roebuck and Co.*

Some Phantom pieces: There's also a back half to the mask, the purple hood part. The rifle barrel is often broken off. *Courtesy of Play With This.*

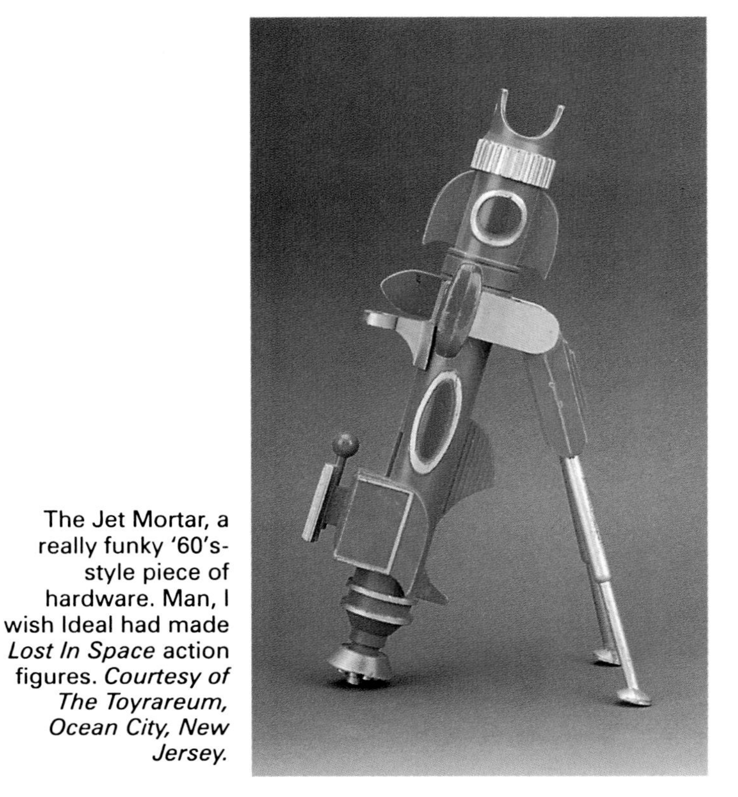

The Jet Mortar, a really funky '60's-style piece of hardware. Man, I wish Ideal had made *Lost In Space* action figures. *Courtesy of The Toyrareum, Ocean City, New Jersey.*

Here's a rarity for you, the shell from the Jet Mortar, shown big enough that you can identify it. *Courtesy of The Toyrareum, Ocean City, New Jersey.*

Action Boy, with kicky beret, the kind you find at an antique show. *Courtesy of The Toyrareum, Ocean City, New Jersey.*

Superman		
Superman Outfit, first issue	$200	$500
Superman Outfit w/flicker ring	$225	$900
Superman In Pieces		
Leotard w/attached cape	$25	
Mask	$20	
Boots	$20	
Belt	$20	
Krypto	$45	
Phantom Zone Projector	$25	
Kryptonite Rock	$35	
Chains, complete	$50	
Aquaman		
Aquaman Outfit, Complete	$200	$450
Aquaman Outfit w/flicker ring	$225	$550
Aquaman, De-Boned		
Leotard	$30	
Mask	$25	
Flippers	$30	
Belt	$30	
Conch Shell	$25	
Lance sword	$35	
Trident	$60	
Knife	$30	
Captain America		
Captain America Outfit	$225	$650
Captain America Outfit, w/ flicker ring	$250	$750
Captain America Divided		
Leotard	$35	
Mask	$20	
Boots, pair	$25	
Shield	$50	
Gunbelt	$30	
Pistol	$30	
Laser Rifle	$40	

Some of Action Boy's accessories, including Khem the Panther. Copies of Khem abound. The copies have crossbars along the hollow of the stomach, the original does not. *Courtesy of Play With This.*

Parts of Action Boy's Superboy outfit. Note that the cape is not removeable. *Courtesy of Play With This.*

Sgt. Fury		
Sgt. Fury Outfit Complete	$200	$500
Sgt. Fury Debriefed		
Jumpsuit	$35	
Mask	$30	
Boots, pair	$30	
Gun Belt	$30	
Helmet w/chin strap	$30	
Pistol (.45)	$25	
Walkie-Talkie	$25	
Bandoleer	$25	
Grenades (3), each	$10	
Machine Gun	$35	
Spider-Man		
Spider-Man Outfit, Complete	$900	$5500
Spider Man Dissected		
Mask	$70	
Leotard	$50	
Boots	$50	
Belt	$100	
Web Fluid Tank w/hose	$100	
Spider-shaped hook	$100	
Spider Flashlight	$60	
Spider Web plastic "line" w/handle	$100	
The Phantom		
Phantom Outfit complete	$175	$650
Phantom Outfit with Flicker Ring	$200	$750
The Phantom Dissipated		
Leotard	$25	
Mask, Face	$20	
Mask, Hood	$20	
Boots	$25	
Gunbelt	$25	
Pistols (two .45s), each	$20	
Rifle	$40	
Knife	$25	
Brass Knuckle w/Phantom imprint	$75	
Steve Canyon		
Steve Canyon Outfit complete	$200	$550
Steve Canyon w/flicker ring	$250	$650
Steve Canyon Panel-by-panel		
Jumpsuit	$30	
Mask	$25	
Flight Helmet	$25	
Oxygen Mask	$30	
Backpack w/straps	$40	
Cap	$20	
Gunbelt	$20	
Pistol (.45)	$15	
Knife	$20	
Boots	$30	

Flash Gordon		
Flash Gordon Outfit, complete	$175	$450
Flash Gordon Outfit w/flicker ring	$220	$650
Flash Gordon Spaced Out		
Space Suit	$25	
Helmet	$20	
Face Mask	$20	
Belt	$25	
Ray Gun	$20	
Boots	$20	
Oxygen Propellant Gun w/ wand	$35	
Buck Rogers		
Buck Rogers w/flicker ring	$400	$1500
Buck Rogers Space-O-Stuff		
Space Suit	$50	
Helmet	$50	
Space Belt	$50	
Boots, pair	$50	
Jets, pair	$100	
Space Gun	$50	
Space Light	$35	
Canteen	$50	
The Lone Ranger		
Lone Ranger Outfit, Red Shirt	$250	$650
Lone Ranger Outfit, Blue Shirt w/ring	$350	$750
Lone Ranger Round-Up		
Shirt (red)	$30	
Shirt (blue)	$40	
Pants (black)	$30	
Pants (blue)	$40	
Hat	$40	
Mask	$20	
Gunbelt (with detachable buckle)	$35	
Pistols, each	$20	
Boots, pair	$25	
Winchester Rifle	$35	
Tonto		
Tonto Outfit w/flicker ring	$400	$1500
Te Quiero, Tonto!		
Shirt	$30	
Pants	$30	
Moccasins (pair)	$60	
Mask	$35	
Gunbelt	$30	
Head Band	$50	
Pistol	$30	
Knife	$30	
Bow	$75	
Quiver	$50	
Four different arrows, set	$80	
Eagle (with feet attached)	$90	

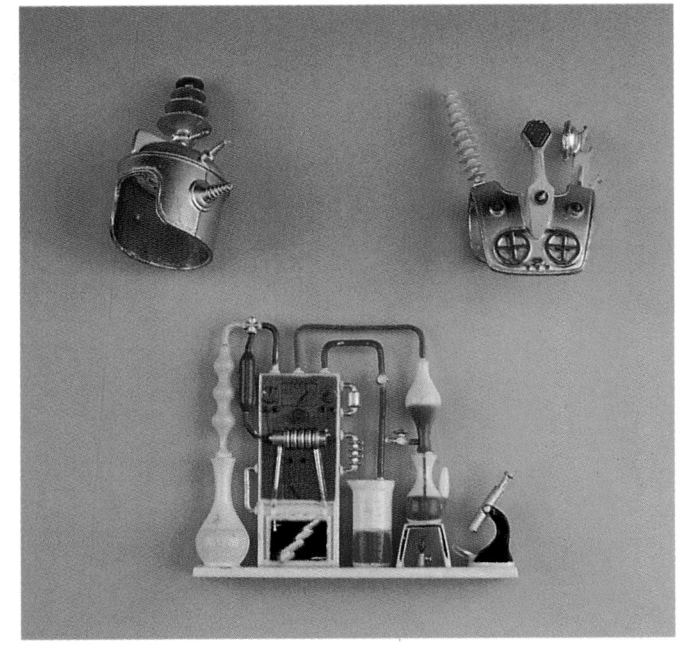

Superboy's translator chest piece and helmet, plus his chemistry lab. He's making a hair restorer for Luthor. *Courtesy of Play With This.*

Some Action Boy Robin stuff. The leggings were attached to the vest (or jerkin) but the cape was separate. *Courtesy of Play With This.*

Some of Action Boy's Aqualad costume parts, including the extremely rare shell axe. *Courtesy of Play With This.*

Captain Action's final and best box, the step-on-Dr. Evil box. *Courtesy of The Toyrareum, Ocean City, New Jersey.*

Green Hornet		
Green Hornet Outfit w/flicker ring	$1200	$4500
Green Hornet Dissected		
Hat	$90	
Mask	$150	
Oxygen Mask	$100	
Scarf	$100	
Coat	$100	
Pants	$75	
Socks, pair	$80	
Shoes, pair	$80	
Stinger Cane	$200	
Gas Gun	$75	
Pocket Watch	$125	
TV Scanner w/phone receiver	$160	
Shoulder Holster	$100	
Action Boy Outfits		
Robin		
Robin Outfit, Complete	$400	$1300
Robin de-feathered		
Mask	$40	
Vest (or Jerkin) w/ leggings	$50	
Cape	$40	
Shoes, pair	$50	
Utility Belt	$50	
Suction Cup Grips (pair)	$60	
Batarang Launcher	$75	
Batarangs, (2) each	$40	
Bat Grenade	$50	
Gloves, pair	$60	

"I'll have my revenge!"

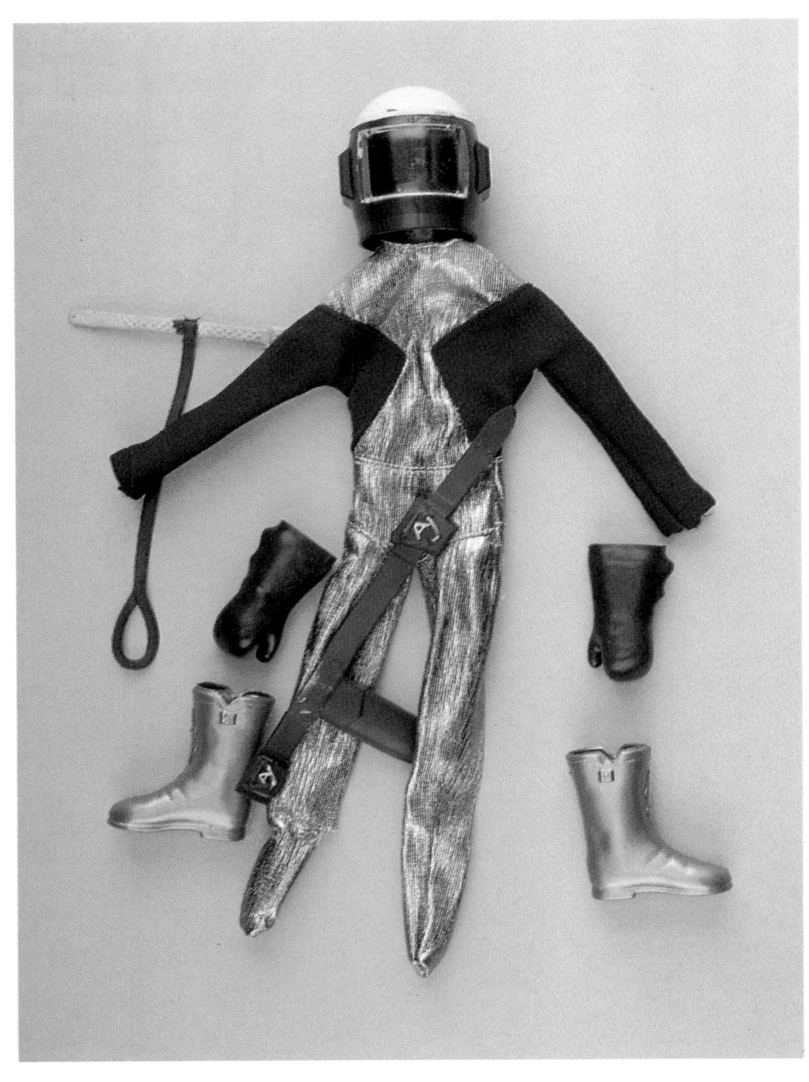

Action Boy's space suit from the photo-box version of same. *Courtesy of The Toyrareum, Ocean City, New Jersey.*

A Dr. Evil Invasion, thanks to Fred Mahn and Paul Levitt for the loan of their Docs.

Superboy		
Superboy Outfit, Complete	$450	$1400
Superboy in Pieces		
Mask	$50	
Leotard w/cape	$60	
Boots, pair	$50	
Belt	$40	
Telepathic Helmet	$100	
Translator Chest Unit	$100	
Chemistry Lab (1 solid piece)	$150	
Aqualad		
Aqualad Outfit, Complete	$350	$900
Aqualad Scaled		
Mask	$40	
Leotard	$40	
Boots, pair	$40	
Belt	$40	
Topo the Octopus	$100	
Sea Horse knife	$45	
Sea Shell Axe	$45	
Swordfish Spear	$80	
Captain Action Accessories		
Directional Communicator	$75	$250
DC Dome Helmet	$25	
DC Chest Pack	$30	
DC Bullhorn	$15	
DC Beam Projector	$20	

Dr. Evil's John Marshall-style mask. Mind you, I think the nose is a definite improvement. *Courtesy of Fred Mahn.*

Power Pack (Jet Pack)	$100	$250
Pack w/handles & thrusters	$50	
PP Gloves, pair	$23	
PP Boots, pair	$15	
PP Flight Helmet	$20	
Weapons Arsenal	$125	$300
(Re-Painted Weapons from Outfit Sets)		
WA Laser Rifle	$20	
WA Laser Pistol	$25	
WA Knife	$20	
WA Revolvers (pair)	$22	
WA Pistol	$15	
WA Machine Gun	$15	
WA Grenades, pair	$30	
WA Storage Rack	$30	
Jet Mortar	$100	$275
JM Mortar Only	$40	
JM Scanner (sight)	$15	
JM Bipod	$15	
JM Shells, pair	$20	
JM Shell Carrier	$20	
Survival Kit	$100	$275
SK Vest	$20	
SK Fishing Kit	$20	
SK Mirror	$20	
SK First Aid Kit	$20	
SK Radio	$20	
SK Folding Spade	$20	
SK 3-piece Extension hook	$25	
SK Machete	$15	
SK Utility Belt	$15	
SK Flare Pistol	$10	
SK Flares (3 in yellow holder)	$25	
SK Dagger	$10	
SK Ammo	$10	
SK Hatchet	$10	
SK Molded Rope	$15	
Parachute Pack	$50	$200
Parachute and Harness	$75	
PP Helmet	$20	
PP Boots	$20	
PP Back Pack	$25	

Dr. Evil's evil outfit and evil, evil gun. He also came with a medallion and sandals, which are virtually impossible to find loose. Err, as this picture proves, I guess. *Courtesy of Play With This.*

Structures & Vehicles		
Quick-Change Chamber Combo Set (Sears)	$900	$1400
Includes Chamber, Cap, & Batman Costume (Quick-Change Chamber Loose)	$300	
Headquarters Carry Case	$300	$900
Dr. Evil Sanctuary	$800	$1300
Chair from Sanctuary	$100	
Silver Streak Car Complete	$800	$1400
Silver Streak w/Garage (Sears)	$1200	$1700
Garage, Loose	$300	
Rockets from Silver Streak, each loose	$60	
Licensed Objects		
Aurora Model Kit	$50	$125
Inflatable, Child-Size Raft / C.A. Logo	$125	$200

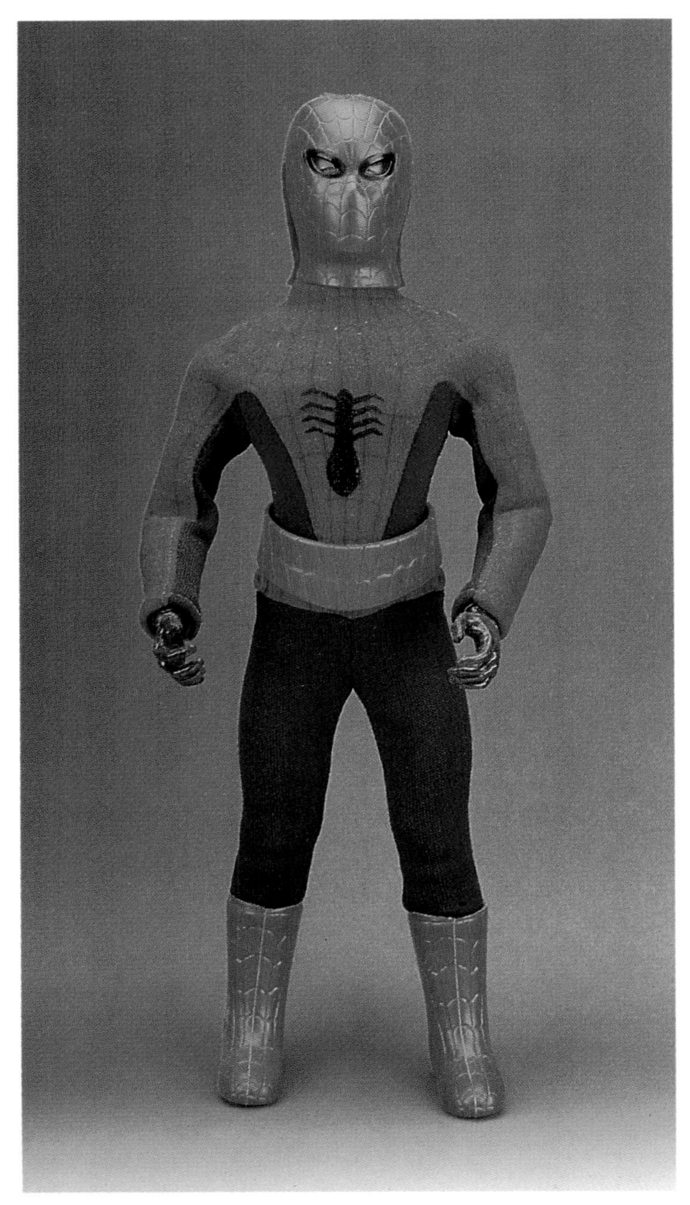

Cap dressed as Spider-Man. The outfit didn't have gloves, so Cap's hands have been painted red. Most characters did not have gloves, even ones like Batman. Well, no toy is perfect. *Courtesy of The Toyrareum, Ocean City, New Jersey.*

Detail of the Spidey belt. *Courtesy of The Toyrareum, Ocean City, New Jersey.*

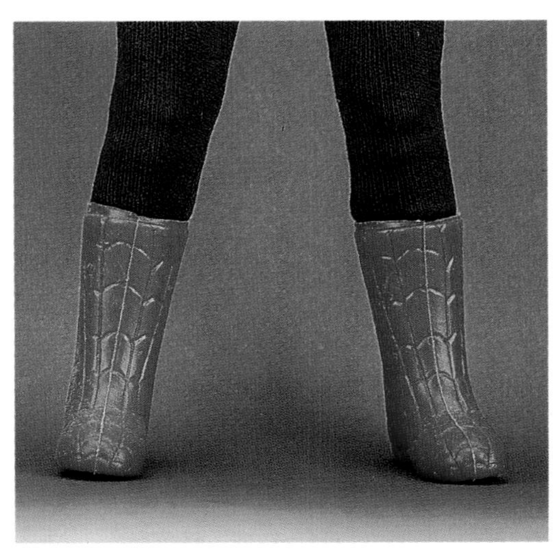

Spidey's boots. *Courtesy of The Toyrareum, Ocean City, New Jersey.*

The spider "familiar" which served as a grappling hook for the spider-line. This is the rare seven-legged version. *Courtesy of The Toyrareum, Ocean City, New Jersey.*

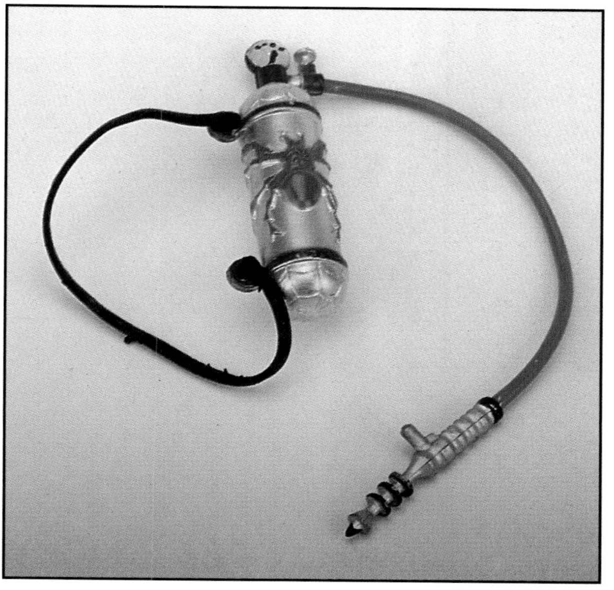

The web fluid tank and spray hose. Of course, we all know that Spidey's shooters were on his wrists, but I guess Ideal felt he needed more accessories. *Courtesy of The Toyrareum, Ocean City, New Jersey.*

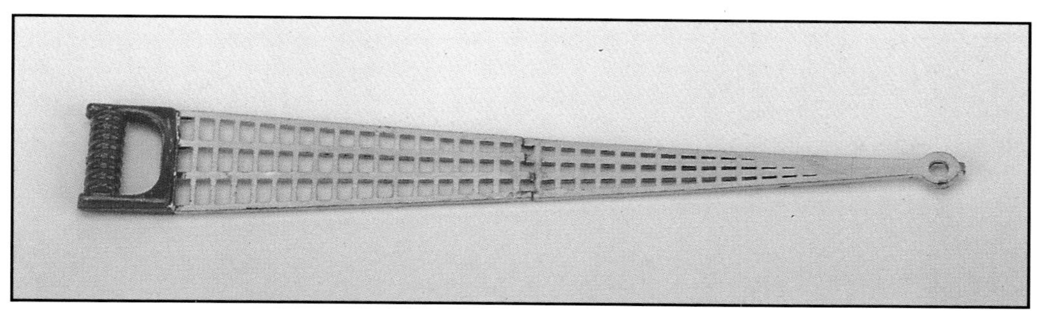

This may look like a web saw but it is supposed to be Spidey's web line, with a handhold. It's even more fragile than it looks, which is why you rarely see an unrepaired one (I mean, one that never needed repairing). *Courtesy of The Toyrareum, Ocean City, New Jersey.*

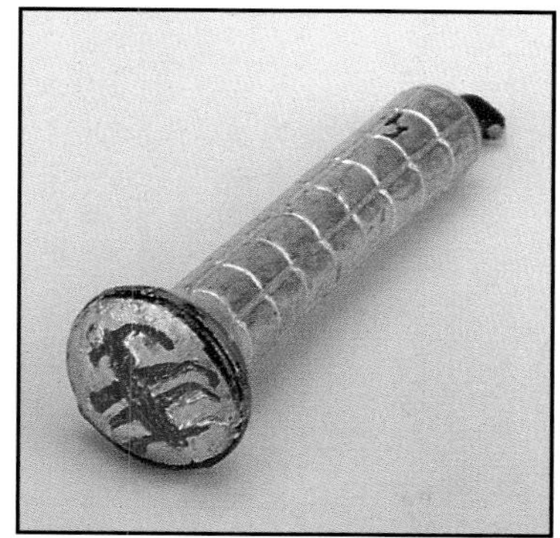

The Spider-Signal, a plastic flashlight. He needs one, trust me. It used to have a hook for clipping onto the belt like Batman's drill and flashlight. *Courtesy of The Toyrareum, Ocean City, New Jersey.*

"Don't get any ideas, Krypto," says Ideal's Super Girl, borrowing a gag from the Introduction. *Courtesy of The Toyrareum, Ocean City, New Jersey.*

Johnny was equally adept at football.
Courtesy of Earl Shores.

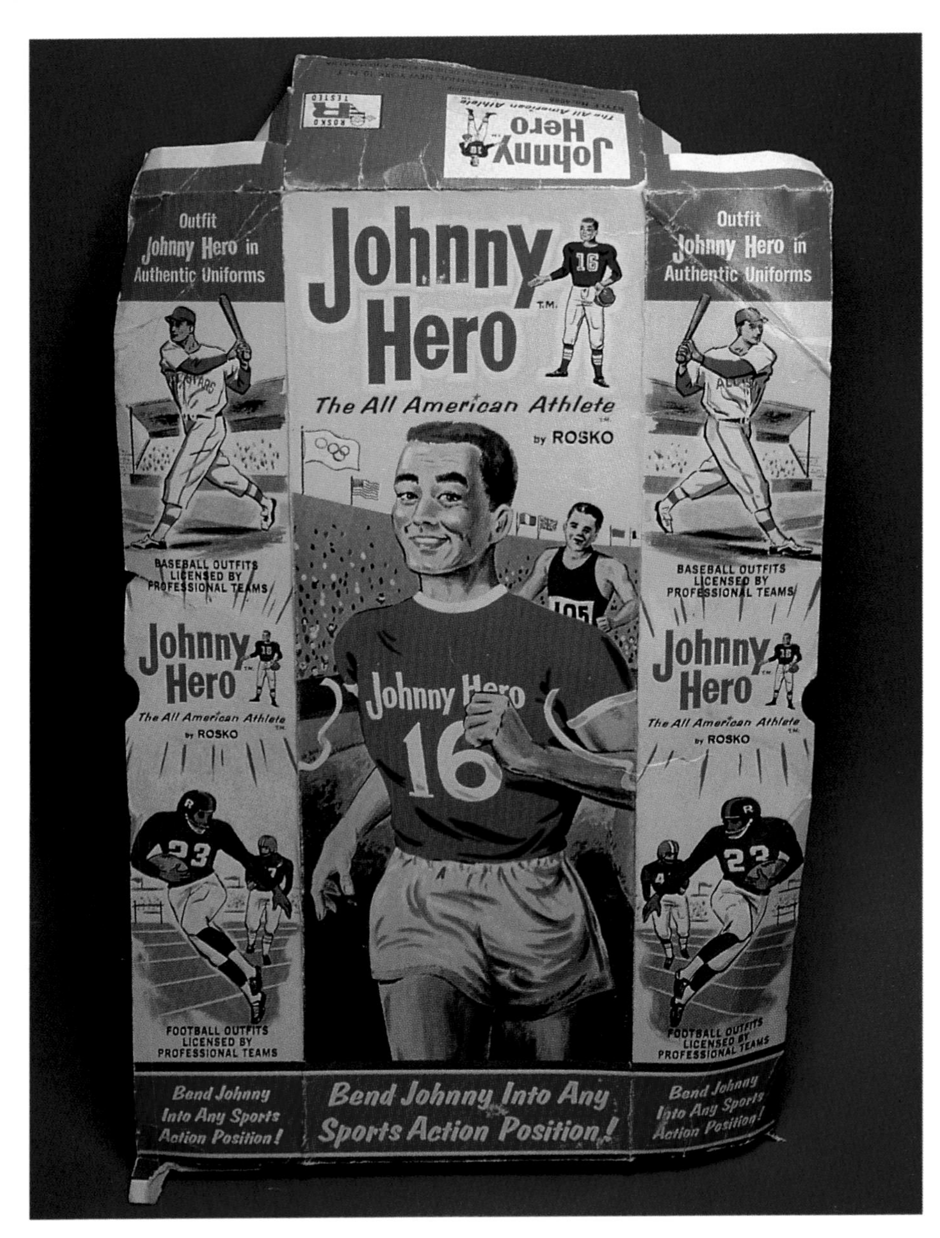

We stepped on this box so you could see the side panels. We love you that much. *Courtesy of Earl Shores.*

Each uniform set consisted of authentically detailed items which included jerseys, pants, appropriate headgear, footgear, and equipment. Johnny was issued with a line of 39 different uniforms representing the NFL, AFL, and both sides of Major League Baseball. "Johnny was one of the earliest items licensed by the NFL, which had been formed only a few years earlier," says Earl Shores, who is the nation's most-recognized expert on Johnny Hero. Shores has made great progress in creating a public awareness about Johnny. You know I never quote anybody but me; however, in this case I simply must defer to the real expert.

Earl has not only made the collecting community more aware of Johnny, but the real world as well, having contributed an article on Johnny to *Sports Illustrated*. Getting info wasn't easy. NFL Properties only had scant information. The figure was available in 1965 and 1966, but then seems to have disappeared for a few years. "That's one of the great mysteries about Johnny Hero," says Shores. "Was he even available a year or two after his premiere? It's hard to believe that an NFL licensed item wouldn't have been a big seller."

True enough, as, in our hurly-burly, hug-a-mug modern world, we are besieged with all sorts of junk associated with NFL and MLB licensing. Since it was (relatively) rare in the late 1960s, all the more reason for eager sports fan kids to have snapped it up back then. You would think.

Many sports stars have long careers, but Johnny vanished after 1966. He resurfaced in 1968 as a tie-in to the '68 Olympics. But there were a few changes. Johnny was now called Johnny Olympic Hero, and he wore a red sweatsuit, and his box came in two alternate versions: blue or red, with white lettering. The uniforms were also repackaged with the "Olympic" logo too, but now they were not in the same league as the original releases. The iron-on numbers on the jerseys were gone, as were the shoulder and thigh pads from the football uniforms.

But the major problem was that these uniform "sets" didn't match up—a Dodgers cap would be included with a Phillies uniform, or even a Mets cap with a Yankees uniform—now there's a fight just waiting to happen! And some packets actually MIXED football and baseball pieces together!

Part of Johnny's Phillies uniform.
Courtesy of Fred Mahn.

I'm not sure about the helmet, but the rest of the stuff is Johnny Hero football equipment.
Courtesy of Fred Mahn.

Worst of all, the reissued uniforms were sold in baggie packs instead of the attractive boxes they originally came in.

And then, Johnny was a gonny. "It wasn't a big marketing push," says Earl Shores. "It was just something that was out there."

It's ironic that such a well-made item wasn't a smash hit, considering that lesser fare (by comparison) is super-popular today. "I think the current 12" Starting Lineups don't measure up to Johnny Hero," says Earl. "The uniform quality is not up to what they did in 1965."

Shores is surprised by the disparity. "Considering that the Starting Line Up series is considered more of a collectible than a toy, you'd think it would be made with a higher quality. Johnny Hero had super detail and quality materials, and yet he was made to be played with in the dirt—and was."

There are a couple of factors that may have contributed to Johnny's, ahh, what's the opposite of longevity? Shortevity? One was, and still is, the difficulty in tracking down a complete uniform. The Johnny Hero uniforms that can still be found in the original packages are usually the crazy mixed-up "Olympic" ones. Collectors seek out whatever they can get, hoping to complete the uniform eventually. The original uniforms in 1965 packaging were issued complete, but they're extremely rare, with a ratio of about thirty-to-one in favor of the later, wacky packages.

But the uniforms aren't the worst problem. The figure's foam latex body, even in the best-case scenario, dries out and degenerates. The latex hardens up so that it splits when you try to bend the arms and legs. And then there's the problem of, well, rot.

Says Earl Shores, "The body dries out when its been exposed to the air. If you find one who has never had his sweatsuit off, or at least was stored away with his sweatsuit on for the last 30 years, then he might be slightly discolored, but his foam will still be in good shape." A common problem is that Johnny figures have often spent the interim years dressed, but shoeless and sockless, so that, "Their feet," as Earl says, "look like they spent a winter at the South Pole without any boots on."

Since Johnny Hero is only just becoming recognized, collecting him is quite a chore. He looks like an imitation G.I.Joe, and his accessories can be confused with the relatively slow-selling Big Jim sports equipment. So, if you're interested in amassing a set of Johnny Hero stuff, you have your work cut out for you.

Mismatched outfits, rotting flesh, and the blank stares of toy dealers who have no idea what you're talking about are the hurdles awaiting Johnny Hero collectors. I wish all Johnny Hero collectors luck, for truly, as these shots of Chuck Eckles' collection prove, a lineup of dressed Johnny Hero figures is a wonder to behold.

Still, there is an upside. Johnny Hero is one of the few under exploited segments of the 1960s action figure pantheon, so pieces can be had for a fraction of what similar G.I.Joe or Captain Action pieces might cost.

Johnny Hero may have been ahead of his time in the 1960s, but it can truly be said that he was a pioneer in the now-successful world of sports action figures.

Plus, he had a really great name.

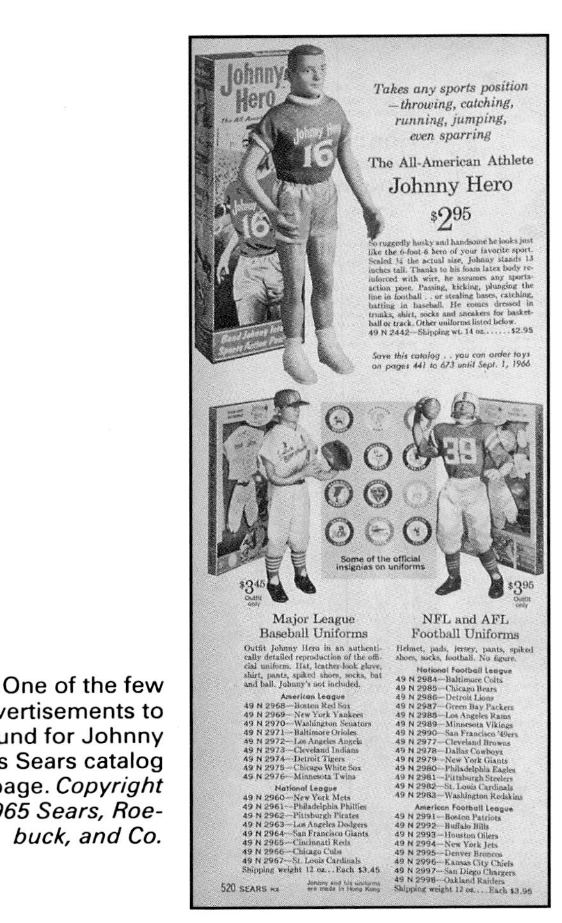

One of the few advertisements to be found for Johnny is this Sears catalog page. *Copyright 1965 Sears, Roebuck, and Co.*

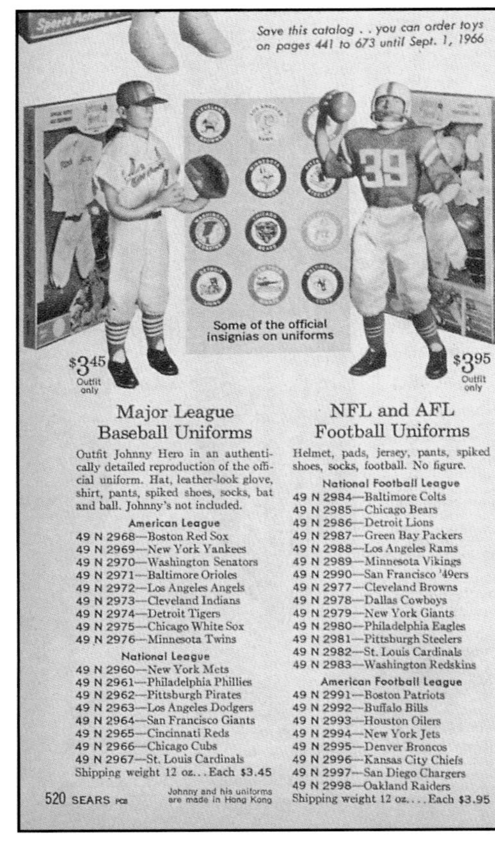

A close-up of the lower section shows a list of his uniforms available through Sears. *Copyright 1965 Sears, Roebuck and Co.*

Johnny Hero Price Guide

	LMC	MIP
Johnny Hero, Complete in 1965 Box	$75	$125
Johnny Olympic Hero (Red Box)	$75	$80
Johnny Olympic Hero (Blue Box)	$75	$100

Uniforms

Mint In Package denotes original boxed versions, since its safe to assume that the baggie packed outfits are not complete anyway. Baggies generally run $30-$50, with teams having relative values corresponding with the boxed prices listed below.

Baseball Uniforms	**LMC**	**MIP**
American League		
Boston Red Sox	$40	$95
New York Yankees	$50	$125
Washington Senators	$50	$95
Baltimore Orioles	$40	$75
Los Angeles Angels	$40	$75
Cleveland Indians	$40	$75
Detroit Tigers	$40	$75
Chicago White Sox	$40	$65
Minnesota Twins	$40	$65
National League		
New York Mets	$40	$125
Philadelphia Phillies	$40	$125
Pittsburgh Pirates	$40	$75
Los Angeles Dodgers	$50	$100
San Francisco Giants	$50	$95
Cincinnati Reds	$50	$75
Chicago Cubs	$50	$125
St. Louis Cardinals	$50	$95
Football Uniforms		
National Football League		
Baltimore Colts	$50	$75
Chicago Bears	$75	$125
Detroit Lions	$50	$95
Green Bay Packers	$75	$125
Los Angeles Rams	$64	$99
Minnesota Vikings	$50	$95
San Francisco 49ers	$65	$100
Cleveland Browns	$50	$96
Dallas Cowboys	$75	$125
New York Giants	$65	$100
Philadelphia Eagles	$75	$125
Pittsburgh Steelers	$75	$125
St. Louis Cardinals	$60	$90
Washington Redskins	$50	$75
American Football League		
Boston Patriots	$65	$100
Buffalo Bills	$65	$100
Houston Oilers	$50	$75
New York Jets	$65	$100
Denver Broncos	$65	$125
Kansas City Chiefs	$40	$100
San Diego Chargers	$40	$125
Oakland Raiders	$75	$125

Here's a typical boxed baseball uniform. *Courtesy of Earl Shores.*

And here's a typical—wait for it!—boxed football uniform. *Courtesy of Earl Shores.*

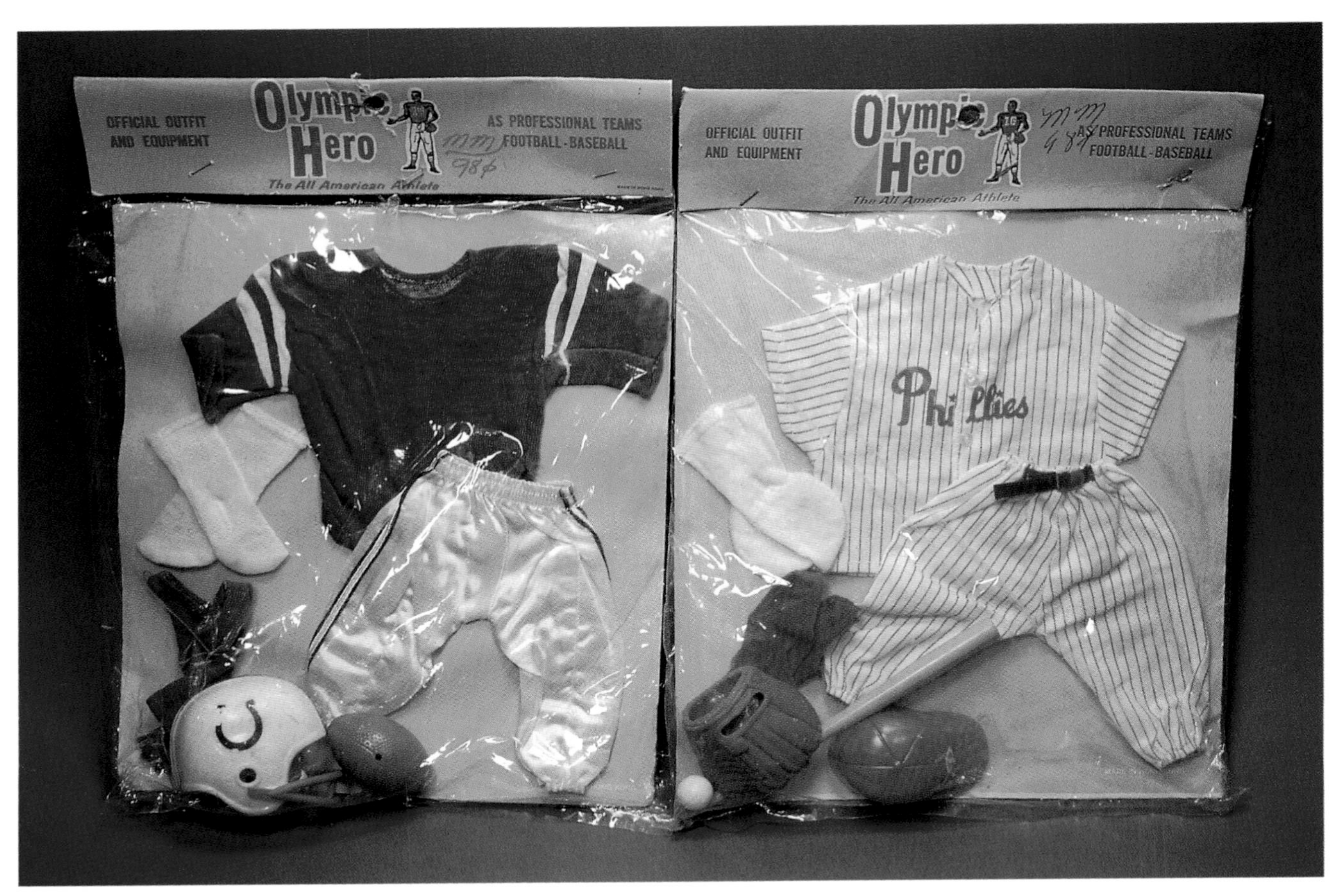

Here are some "baggie" pack uniforms from the Olympic Hero line. *Courtesy of Earl Shores.*

The Chuck Eckles collection.
Courtesy of Chuck Eckles.

The Chuck Eckles collection.
Courtesy of Chuck Eckles.

The Chuck Eckles collection.
Courtesy of Chuck Eckles.

The Chuck Eckles collection. *Courtesy of Chuck Eckles.*

The Chuck Eckles collection. *Courtesy of Chuck Eckles.*

The Chuck Eckles collection. *Courtesy of Chuck Eckles. All photos in this chapter by Earl Shores.*

Chapter Seven

Kiss, Kiss, Buy, Buy: The Super-Spies!

Copyright 1965 Sears, Roebuck, and Co.

The world of espionage is a world populated by people with a license to kill—a world of dedicated professionals (along with the occasional talented amateur). At least, that's what the movies and TV would have us believe.

In the early 1960s, the subject of spies was coming out of the woodwork and they were the focus of national attention—the opposite of what the average spy wants, of course. When President John F. Kennedy announced that *From Russia With Love* was one of his favorite books, all the white bread Kennedy wannabes started buying Ian Fleming's novels. Everybody was getting spy crazy.

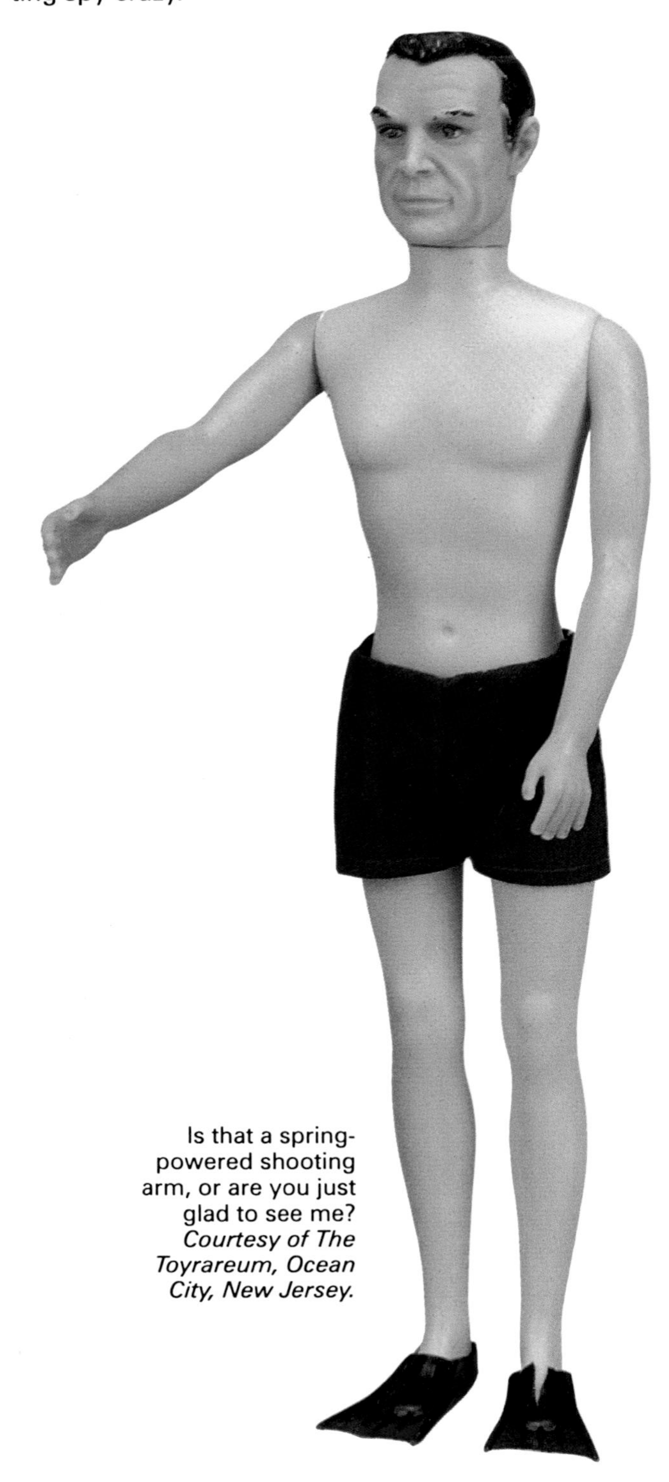
Is that a spring-powered shooting arm, or are you just glad to see me? *Courtesy of The Toyrareum, Ocean City, New Jersey.*

But the real mania started when the first James Bond movie, *Dr. No*, opened our eyes to a whole new kind of mind candy. When Ursula Andress stepped out of the surf and onto the shores of Crab Key, the public's interest in spy stuff escalated to stratospheric proportions. From that moment on, the spy became a glamorous new kind of super hero, and a merchandising icon worldwide.

It wasn't lost on the TV crowd, either. James Bond rip-offs showed up on the small screen, both serious (*The Man From U.N.C.L.E.*) and silly (*Get Smart!*). By the time the James Bond merchandising boom hit, everybody was making, or planning to make, imitation Bond movies. Dean Martin was the so-so Matt Helm, James Coburn starred in two unwatchable Flint movies, and even Sean Connery's brother starred, alongside *From Russia's* Daniela Bianchi and *Thunderball's* Adolfo Celi, in the medium-budget Italian flick *Operation Kid Brother*, also known as *Operation Double 007*. To get in the mood for this chapter, at lunch today I watched half of *Kid Brother* while absorbing a box of Devil Dogs. Then, baby, I was ready!!!

Anyway, going back to the early Sixties, suddenly, kids went from playing cowboys and Indians to pretending to be agents and assassins. And, of course, the toy companies were ready for them. Spy toys of all kinds could fill a book by themselves. Someday I'll write one, no doubt, but for now let's take a look at the most exciting kinds of spy toys, the action figures!

It was with the release of the James Bond movie *Goldfinger*, the third in the series, that Bond merchandising really began. Both *Dr. No* and *From Russia With Love* had been adult-oriented spy adventures, but with *Goldfinger*, the Bond films came more into the domain of the family market, with their colorful villains and amazing gadgets. The first figural item from *Goldfinger* was a model kit produced by Airfix. This pitted James Bond against Oddjob in a diorama scene.

But with the release of *Thunderball*, and in the wake of the successful debut of G.I.Joe, proper James Bond action figures were issued. A.C. Gilbert was never a big player in the action figure market, preferring instead to concentrate on the erection sets which made them famous and paid their bills. Even so, they snagged the first really killer licensing deal in action figure history. Gilbert released 12" action figures based on several different licenses.

The first and most desirable figures were from the James Bond series. The figures that were produced, as the accompanying catalog pages show, included James Bond, Oddjob, and a series of accessories. What you may not be able to tell from the photos is that the dolls are practically immobile. Whereas G.I.Joe had 21 moveable joints, and Johnny West had over a dozen, Gilbert's spy guys had only five: the neck, both shoulders, and both hips. It was pretty pathetic! These stiff figures made for some pretty difficult play, having James Bond lurch around like Frankenstein's monster, his legs stiff, his arms held straight out in front of him. Oddjob fared even worse, sculpted permanently into the crouching position he assumed when he threw his hat. He did however have two spring-powered arms, the one that threw the hat and the other which gave a karate chop.

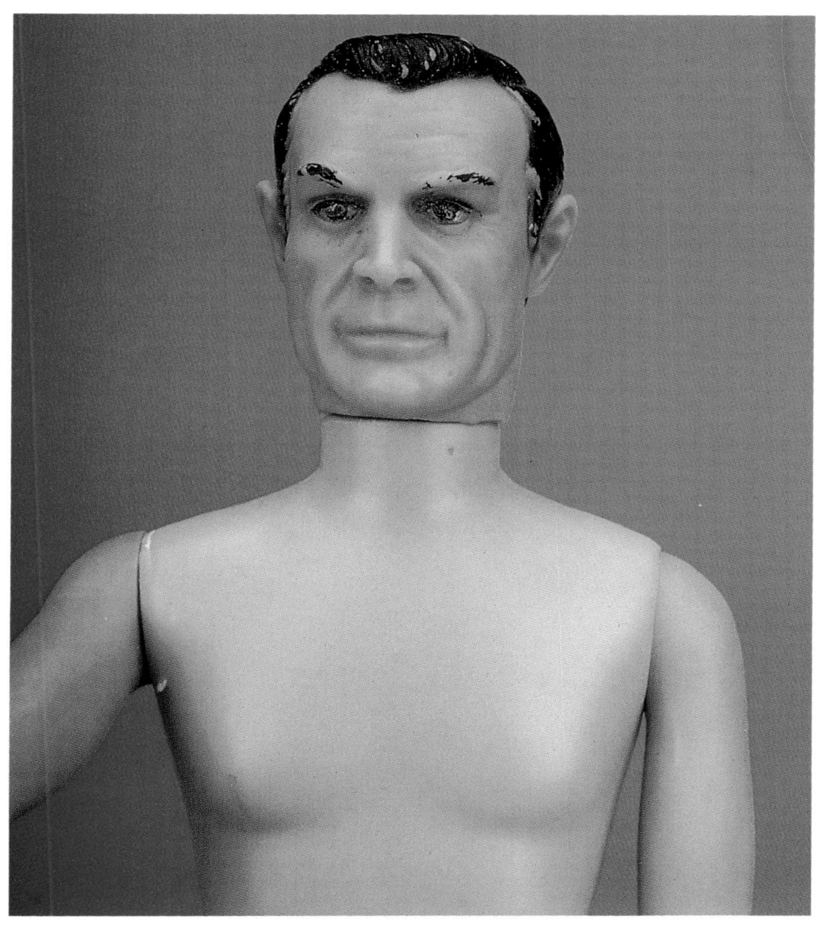

Geez, Sean, you got gas or something? *Courtesy of The Toyrareum, Ocean City, New Jersey.*

James Bond's arm was also spring powered. Somehow, I've never figured out how, when his arm sprang up, it supposedly activated the cap-firing metal pistol he came with. Those pistols are ridiculously hard to find, even in boxed figures which have been previously opened.

These figures, as you can see, were originally sold exclusively through Sears. Bond and Oddjob supposedly came dressed in formal attire, although I've never seen any that actually were. The problem in researching these toys is that most spy fans prefer stuff like the kid-sized attaché cases, so the realm we're discussing here has a ways to go before it's fully explored. I'll do my best to cover everything, though. Gee, I hope you bought the book *before* you read that!

A year or so after the line came out, it went nationwide. Reportedly, Ideal took it over from Gilbert at about this time. The James Bond figure came boxed in a sport shirt and swimsuit combination. Oddjob still had his hat, but wore an inexpensive karate outfit.

Accessories in a bewildering variety were released for James Bond. These included a Disguise Kit complete with trench coat, a Scuba set with a working spear gun, and a *Thunderball* Set. This *T-ball* Set came complete with the jet pack as seen in the *Thunderball* teaser sequence. Over in England, kids enjoyed additional sets: a Ski Suit, a Raincoat Outfit, and a black Tuxedo Suit like the one that the Sears version of the Bond doll supposedly wore.

Gilbert also released a series of 3" painted figurines. These figurines look very good in the catalog paintings but, as you can see from the photos, ain't too impressive in reality.

Also in the figural category are Gilbert's two heavy-rubber hand puppets. These are full body figures of Bond and Oddjob that stand up by themselves and are a valued part of any James Bond action figure collection.

And speaking of Bond and Oddjob, like I have been for the whole chapter, Aurora eventually got around to releasing figures of them in their model kit series. There were packaged separately, but could be put together in a battle scene not dissimilar to Airfix' 1964 figurines.

It was during this period that Corgi was going nuts and issuing vehicles based on all kinds of licenses. Spies and James Bond in particular were certainly not ignored. But one Corgi James Bond item stands out in particu-

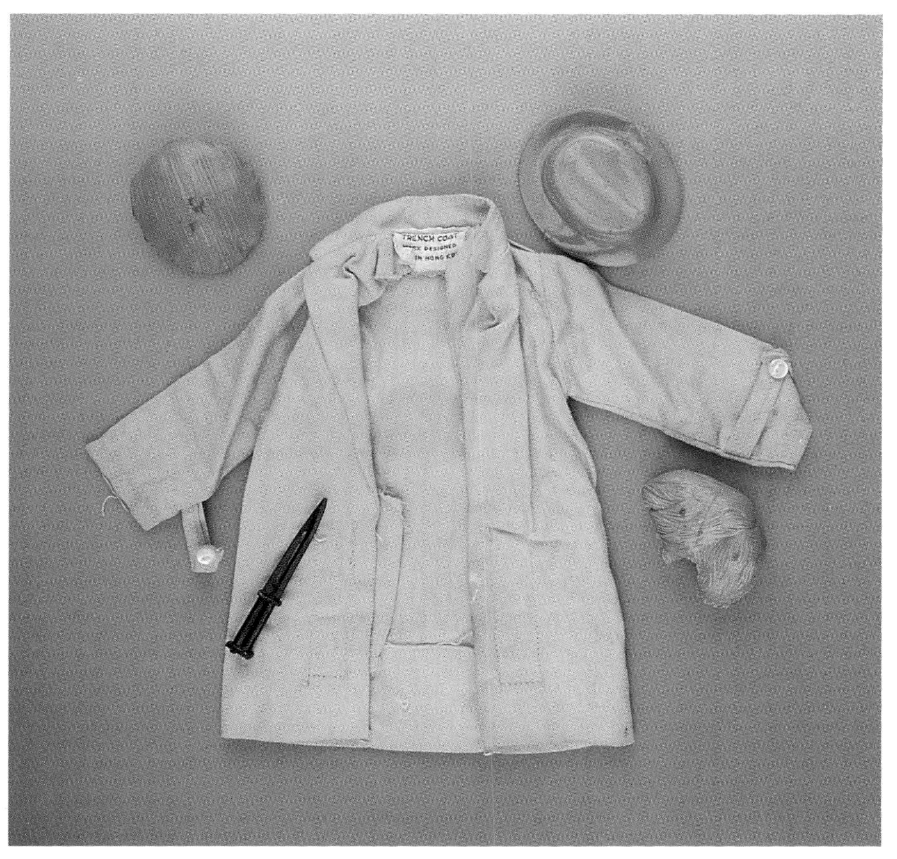

Bond's trench coat and several of what look suspiciously like Mike Hazard accessories. *Courtesy of Play With This.*

Oddjob prepares to throw his deadly derby. *Courtesy of The Toyrareum, Ocean City, New Jersey.*

lar, the one with actual characters from a movie included with it. This is the sports car driven by Aki in the film *You Only Live Twice*. This is my personal favorite Bond film because of its plot line, Ken Adam's design for the volcano base, and the co-stars.

You see, unlike many of Sean Connery's early co-stars, his *Twice* flirtations were big names in their own homelands. Karin Dor, as the S.P.E.C.T.R.E. agent, has made an endless string of films. Mie Hama, as the girl Bond "marries," is a huge star, having top lined *King Kong Versus Godzilla* and played the evil Madame X in Toho's *King Kong Escapes* (the one with the robot Kong). Aki was played by Akiko Wakabayashi, the prophetess of doom in *Ghidrah The Three-Headed Monster*. Wakabayashi's ability to handle English was such that she was removed from her intended role as the bride and she switched parts with Mie Hama. Therefore, it's especially thrilling to me that Corgi's car includes figures of Bond and Aki.

Hama and Wakabayashi also appeared in a spy comedy called *Key Of Keys*, alongside Toho's other femme fatale, Kumi Mizuno. This film was redubbed and reedited into the infamous Woody Allen farce, *What's Up Tiger Lily?* I have not been able to confirm the rumor that there is a planned remake, a Martin Lawrence vehicle to be called, *Yo! Wassup Tiger Lily?*

All this Bondage insured that the other major spies would turn up in the toy aisles as well. For several

Right: Domino carded. I never knew that Bond had a line after "What sharp little eyes you've got," until I bought *Thunderball* on prerecord. ABC cut out his reply, "Wait'll you get to my teeth" on every broadcast I ever saw. *Courtesy of The Toyrareum, Ocean City, New Jersey.*

Goldfinger mis-packaged on a Moneypenny card. It's Goldpenny! *Courtesy of The Toyrareum, Ocean City, New Jersey.*

The Girl From U.N.C.L.E.		
(Marx)	$600	$900
The Avengers		
Corgi Steed with Bentley	$150	$250
Corgi Emma Peel w/ Sportscar	$200	$300
Corgi Gift Set	$350	$500
Emma Peel Doll (Fairylite)		
Emma Peel Version 1, Fashion Clothes	$900	$1500
Emma Peel Version 2, Action Suit	$1500	$2500
Assorted Spies		
Honey West Doll (Gilbert)	$150	$250
Ocelot Set (Pet Set)	$50	$100
Assorted Accessory Cards, each	$25-$35	$50-$60
Mike Hazard Secret Agent (Marx)	$250	$450
Mike Hazard w/ Cardboard Store Display	$600	

Largo carded. "Largo" is Spanish for "long." *Courtesy of The Toyrareum, Ocean City, New Jersey.*

Above: A typical two-pack. *Courtesy of The Toyrareum, Ocean City, New Jersey.*

Left: This card features the Dragon Tank and Largo's yacht, the Disco Volante (Flying Saucer). These vehicles were not in scale with the little figures, obviously. *Courtesy of The Toyrareum, Ocean City, New Jersey.*

The "M" figure loose. It's plastiriffic! (John Marshall Language! Not real English, but an amazing bastardization!) *Courtesy of Play With This.*

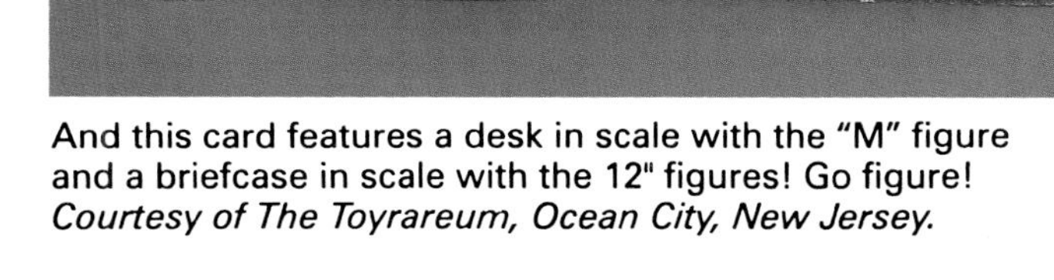

And this card features a desk in scale with the "M" figure and a briefcase in scale with the 12" figures! Go figure! *Courtesy of The Toyrareum, Ocean City, New Jersey.*

Above: This ten-figure gift set looks really cool in the box, doesn't it? You bet it does. Kiss me, you fool. *Courtesy of The Toyrareum, Ocean City, New Jersey.*

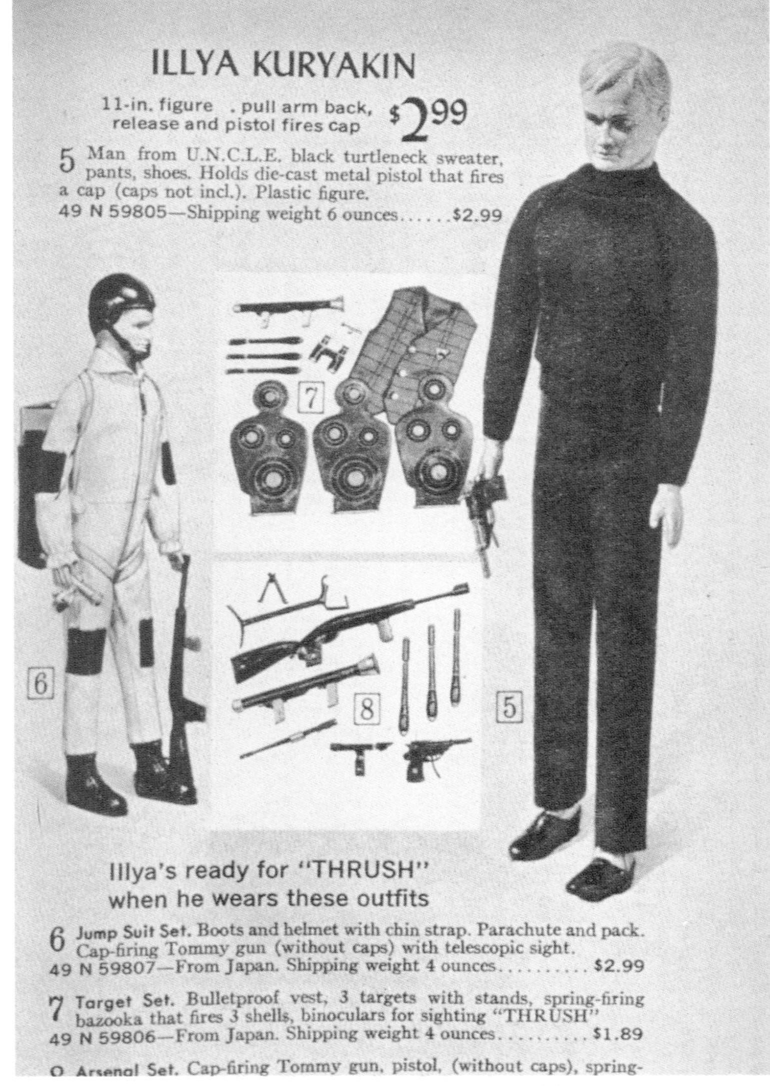

ILLYA KURYAKIN

11-in. figure . pull arm back, release and pistol fires cap $2.99

5 Man from U.N.C.L.E. black turtleneck sweater, pants, shoes. Holds die-cast metal pistol that fires a cap (caps not incl.). Plastic figure.
49 N 59805—Shipping weight 6 ounces......$2.99

7 6 8 5

Illya's ready for "THRUSH" when he wears these outfits

6 Jump Suit Set. Boots and helmet with chin strap. Parachute and pack. Cap-firing Tommy gun (without caps) with telescopic sight.
49 N 59807—From Japan. Shipping weight 4 ounces.......... $2.99

7 Target Set. Bulletproof vest, 3 targets with stands, spring-firing bazooka that fires 3 shells, binoculars for sighting "THRUSH"
49 N 59806—From Japan. Shipping weight 4 ounces.......... $1.89

8 Arsenal Set. Cap-firing Tommy gun, pistol, (without caps), spring-

Left: Illya Kuryakin as sold by Sears. *Copyright 1966 Sears, Roebuck, and Co.*

One of the big mistakes Gilbert made was in their packaging. Real colorful and eye-catching, huh? ... I'm sorry, I had a little sarcoleptic (def.: an attack of sarcasm) fit just now. *Courtesy of The Toyrareum, Ocean City, New Jersey.*

Here he is out of the box. Looks like a lily to me! *Courtesy of The Toyrareum, Ocean City, New Jersey.*

"Friend, good!" Look at this version's head. It's a good example of how the same production line can yield different quality items. *Courtesy of Play With This.*

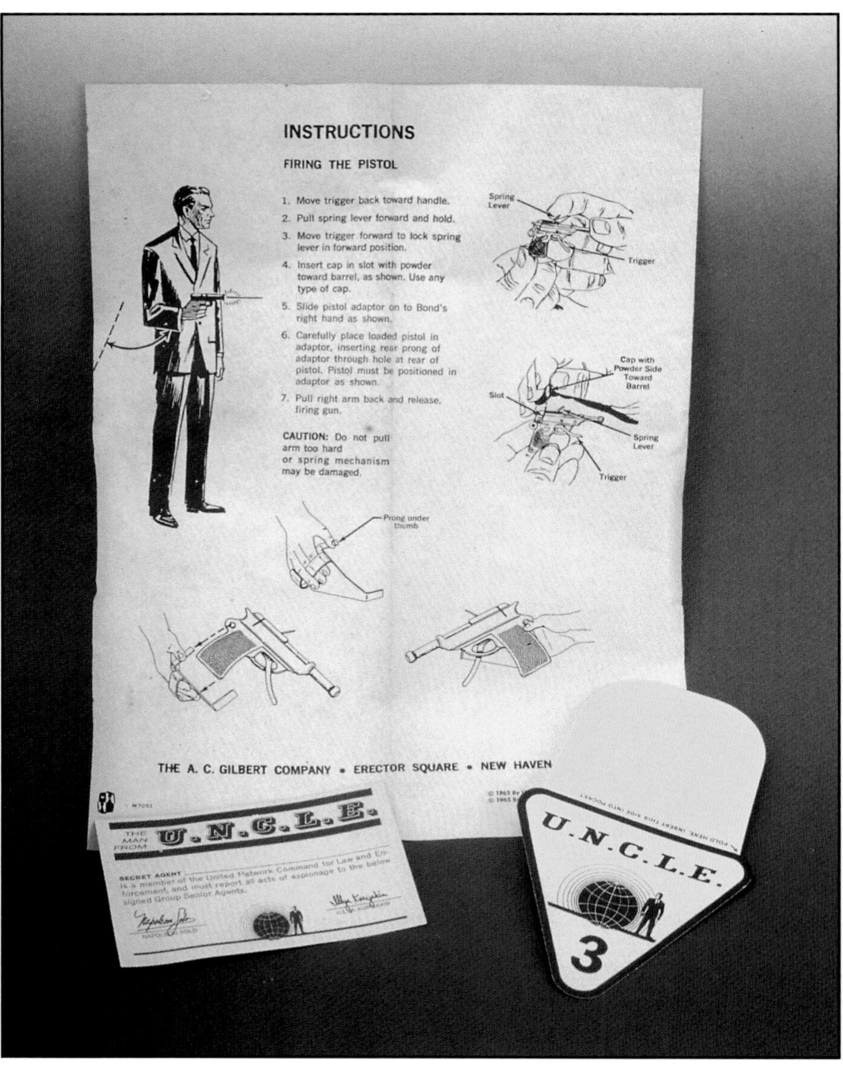

The paperwork from the U.N.C.L.E. dolls. More cool stuff you never see in other books! *Courtesy of The Toyrareum, Ocean City, New Jersey.*

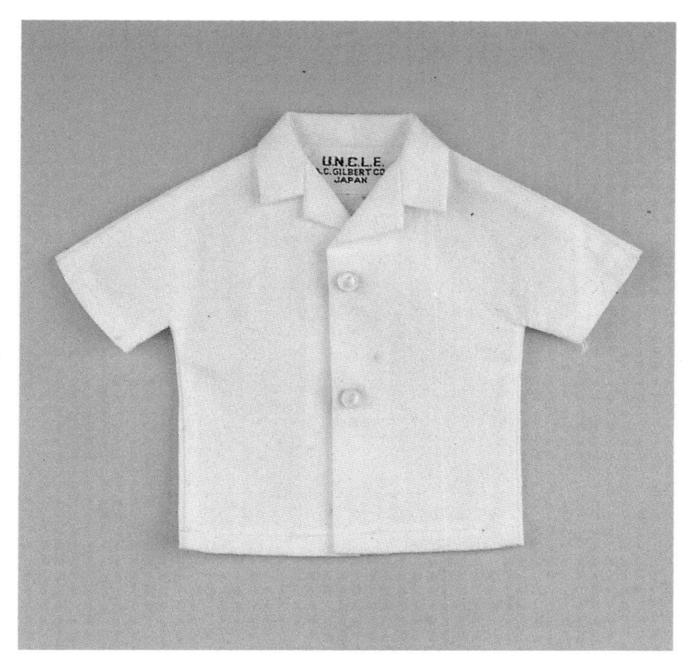

An U.N.C.L.E. shirt from an accessory set. Look! It's white! *Courtesy of Fred Mahn.*

An accessory card for Honey West, TV's Private Eye-Full (ouch). *Courtesy of The Toyrareum, Ocean City, New Jersey.*

One of the U.N.C.L.E. figures from Marx, although the gun barrel is off. *Courtesy of Play With This.*

Chapter Eight

The Amazing Mind Of Irwin Allen

Copyright 1966 Sears, Roebuck and Co.

Mr. Irwin Allen was a film and television producer who will always be remembered for his contributions to two major genres. One genre he left his mark on was the 1970s disaster picture, a genre he virtually created. By "disasters" I don't mean flicks like *Congo* or *Blues Brothers 2000*, but pictures dealing with life-threatening forces beyond man's control. Such films as *The Poseidon Adventure, The Swarm,* and *The Towering Inferno* boasted enormous budgets and all-star casts designed to appeal to all audience age groups.

The other arena he conquered was fantastic television. Irwin Allen was responsible for hundreds of hours of the most amazing adventures that audiences of the 1960s had seen. And viewers experienced all the action and danger right in their living rooms.

These TV shows begat numerous toys. And that's where we come in. But first, a little TV history. The Fall 1964-1965 TV season saw the debut of a television series based on Irwin Allen's successful film, *Voyage To The Bottom Of The Sea*. Actor David Hedison, the original Fly, was asked to portray sub captain Lee Crane. Not having recovered from his participation in Irwin Allen's 1960 remake of *The Lost World*, Hedison hemmed and hawed and had a hell of a headache while trying to decide whether to sign up. (He had, in fact, turned down the film version.) After all, despite its production values and all-star cast, Hedison recalled that *The Lost World* had this air of, well, stupidity about it. The *Voyage* TV show promised to be more of the same, except with a weekly paycheck attached.

But when Hedison learned that veteran actor Richard Basehart had signed to play Admiral Harriman Nelson, he figured, hey, what was good enough for Basehart was good enough for him. In fact, Richard Basehart was no stranger to sea adventures, having played Ishmael in the Gregory Peck-starring version of *Moby Dick*.

The *Voyage* playset, first version. *Copyright 1965 Sears, Roebuck, and Co.*

The *Voyage* playset, second version. For a buck more you got a cool giant whale, as seen in *The Ghost Of Moby Dick, Jonah and the Whale, The Shape Of Doom*, and as stock footage in other episodes. *Copyright 1966 Sears, Roebuck, and Co.*

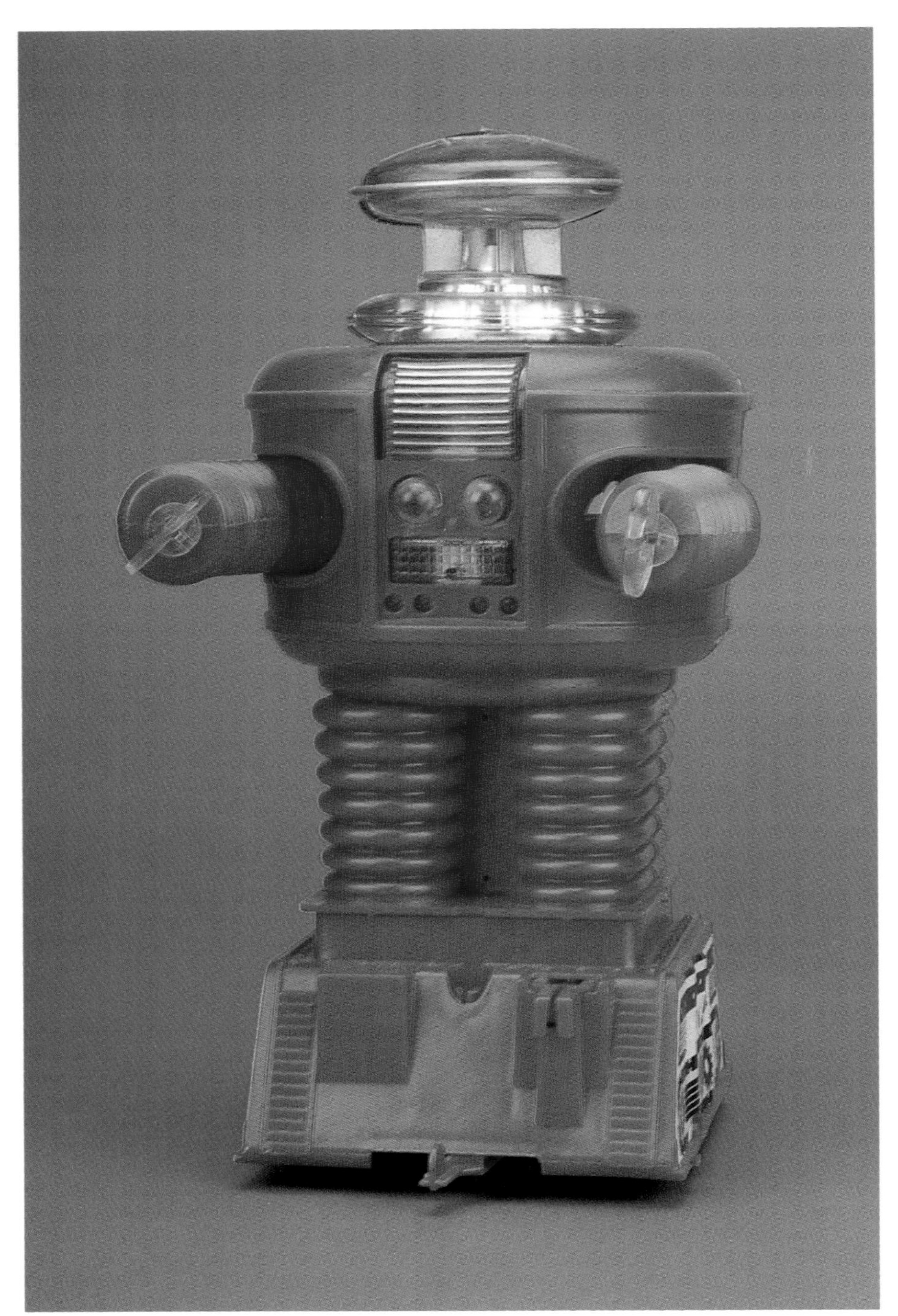

Here's the second-greatest *Lost In Space* toy of the 1960s: The Remco Robot! He was usually molded in two colors, typically blue and red. But here's a rare all-red version. *Courtesy of The Toyrareum, Ocean City, New Jersey.*

The TV series of *Voyage* began its run in black and white, and the tone of the stories was usually appropriate to the color palette. Everything was serious, as the crew of the Seaview matched wits with enemy saboteurs and had James Bond style adventures.

There had, however, been fantasy elements in the original *Voyage* film. There was an attack by a giant squid and, later, a super-giant octopus. Not surprisingly, the footage of these beasts was eventually incorporated into the wilder episodes of the series. Richard Basehart encountered another monster whale, a gray one this time, in the episode *The Ghost Of Moby Dick*, which co-starred June Lockhart, the mother from Lassie. Other fantastic beasts crept in as well. Leslie Nielsen stopped by, pursuing a giant manta ray in *The Creature*. Nick Adams and Yvonne Craig tried to escape an island populated with dinosaur footage from Allen's *The Lost World*, in the episode *Turn Back The Clock*. "Here's a hot one for ya, Lizard Lips, straight from Yancy Street!" And, ultimately, in *The Condemned*, the Seaview crew encountered their first, born-for-TV, bonafide monster: a giant stack of seaweed topped with two heads, each consisting mostly of a glowing, bulbous eye.

It was the appearance of this creature (which I'll nickname LuluBelle after a famous two-headed calf) that established the monster-oriented tone of *Voyage* in its second season. LuluBelle would return several times on *Voyage*, in both new appearances and through stock footage.

Near the start of the following 1965-1966 TV season, the Sears Christmas catalog came out, featuring a Remco Voyage To The Bottom Of The Sea Playset. It came with a massive 18 inch long Seaview submarine, molded for some reason in yellow (the "real" one is blue).

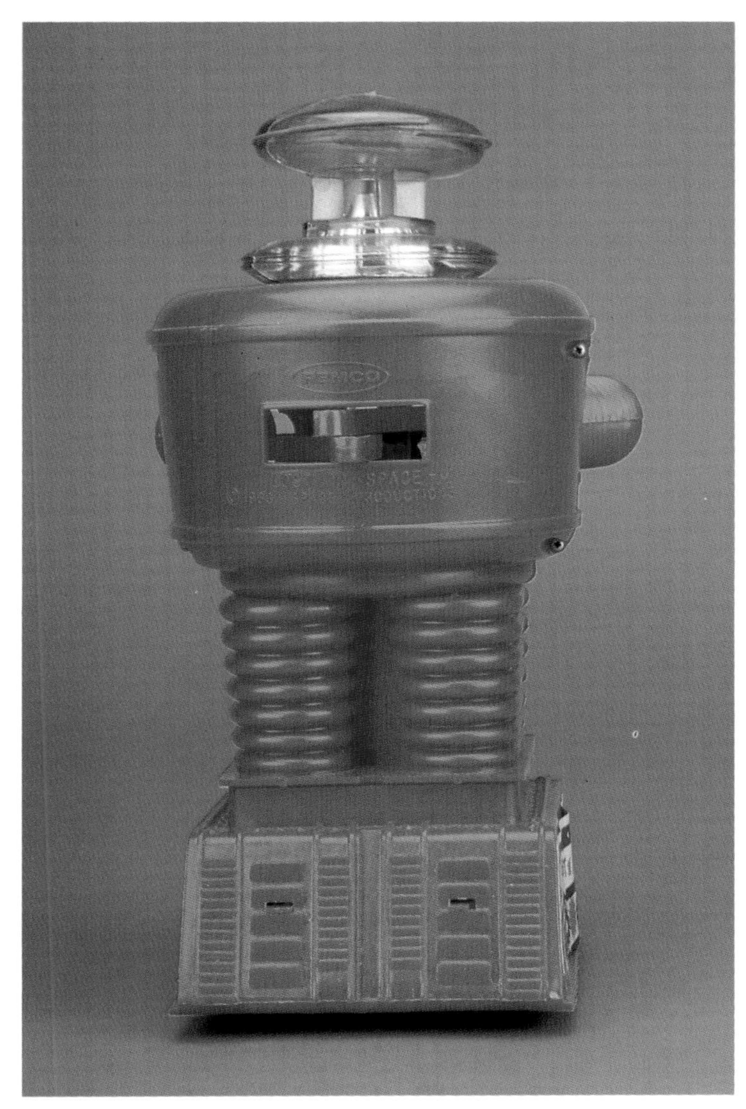

We thought that, with something this rare, you should see as much of it as possible, so you can identify all the parts you need. Here's the back, showing the arm control levers. All detail shots are of a Robot that comes to us *Courtesy of The Toyrareum, Ocean City, New Jersey.*

We bring you the head of the Mechanical Man.

Here is the bottom, showing the bar that held the batteries in place.

It also came with two bubble-topped mini-vehicles that resembled nothing seen on the show. But thanks in part to the performance turned in by LuluBelle and the other giant creatures, two monsters were also included in Remco's playset. One was a giant octopus and the other was kind of a spider-crab fellow. This charmer also appeared as an ally of the giant bug, Horrible Hamilton, in Remco's Horrible Hamilton bugs-vs-soldiers play set series (see Super Freakout Grab Bag).

Meanwhile, on TV, *Voyage* enjoyed its second season, now in color. There was a wisecracking new character, Chief Francis Sharkey, ably played by Terry Becker. There was also an awesome new Flying Sub that allowed the crew greater mobility in the battle against an endless parade of cool 'n kooky monsters and aliens.

That same autumn, *Voyage* was joined by a sister series, whose first season was in black and white. It starred June Lockhart, fresh from her *Voyage* guest appearance, along with Guy Williams, formerly Zorro.

That program was called *Lost In Space*.

Lost had a bit of a rough trip from concept to broadcast. The pilot was filled with exciting enough footage, as The Robinsons, America's first family in space, got sidetracked by a meteor storm and were forced to crash-land on an arid, uncharted planet. Harsh conditions forced them to migrate from pole to pole, encountering the catacombs of a long-dead civilization and gigantic marauding cyclops monsters.

The thing was really impressive. But as the story goes, when Irwin aired the pilot to 20th Century Fox executives, they were laughing at it, because it seemed even cornier than *Voyage* had ever been. It needed retooling.

Two additions were made. One was an awesome Robot, performed by Bob May and voiced by CBS announcer Dick Tufeld. The other was a villain, Dr. Zachary Smith. Allen's production team tapped veteran TV performer Jonathan Harris, who had made

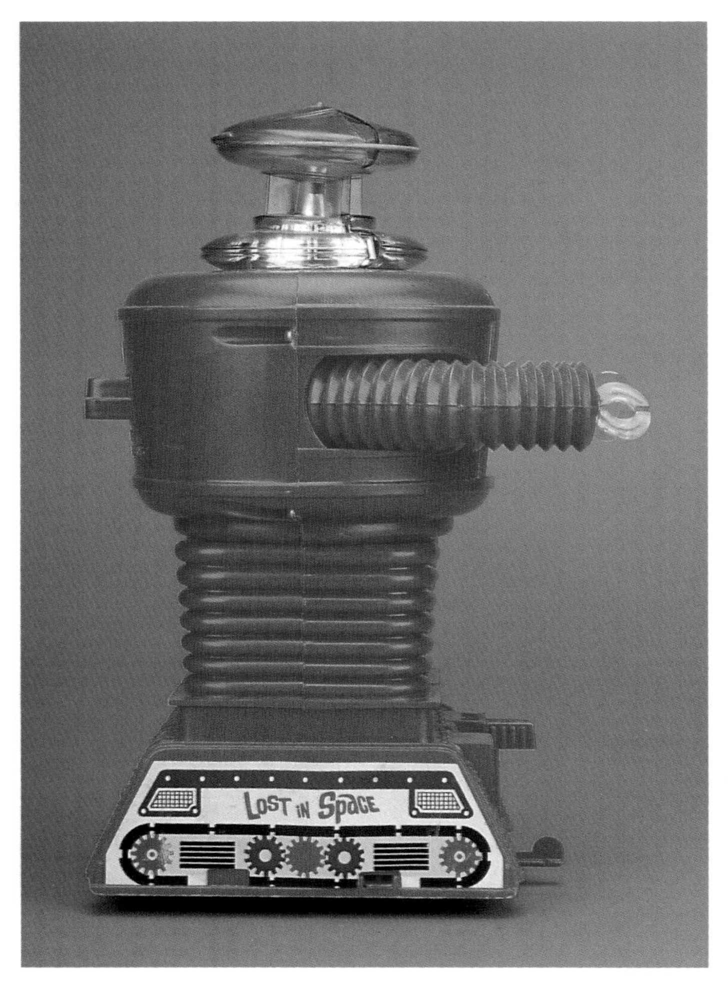

The Robot's right side.

The Robot's left side. Any lesser book would have just reversed the photo, but we give you the real thing!

the arching of eyebrows a veritable art form. Harris quickly realized that if he continued to play the character of Smith as written, as the cold-blooded saboteur that he was in the first few episodes, he surely would be killed off. So he slipped a kind of cowardice and laziness into the character, transforming Smith into an endearing villain, a lovable rascal, and a character upon whom a great deal of my impressionable young personality was molded.

I identified with Smith. He was my Id, portrayed on television. I didn't like to clean up my room. Smith lied his way out of work. I liked to sneak candy before supper. Smith was constantly filching from the food supply. Kids of the 1960s were supposed to identify with young Will Robinson, but he was merely obnoxious. It was Smith with whom the young audience really identified.

Many sources state that there was a great deal of animosity on the set of *Lost In Space* toward Jonathan Harris. He was a late addition and yet was, for all intents and purposes, the star of the show. His "special guest star" status, in fact, was his agent's way of giving him great billing while still having to be listed last in the credits. Heather Locklear took her cue from Jonathan Harris, which is why she, too, is billed last on *Melrose Place*, but as Special Guest Star, every week. I think they should be buried next to each other. "Here lie Jonathan Harris and Heather Locklear, America's Special Guest Stars. They paid their co-stars' rents."

Lost In Space was a substantial hit, holding its own against that national phenomenon *Batman*, which began that January (1966) in color. The constant barrage of monsters helped, including an appearance by LuluBelle as the main antagonist in *The Raft* and a cameo by her in *The Keeper* as well. Things were going great in Irwin Allen land. *Lost* was renewed for a second season (to be broadcast in color), and *Voyage* was renewed for a third.

As those seasons began, Mattel released, exclusively to Sears, a *Lost In Space* Switch 'N Go playset. Switch 'N Go was a system that used air pressure to activate remote control mechanisms ... or something like that. The *Lost In Space* set was by far the most elaborate *Lost In Space* toy ever made, and, fortunately, it included figures of every single cast member, so I can discuss it here.

The figures themselves, although only toy soldier size, were convincing likenesses of all the Robinsons, Don West (molded in a sitting position to drive the Chariot), Dr. Smith, and even Debbie the Bloop! Plus, there was a semi-accurate version of the Robot that could follow the Chariot around. There was a kind of "spaceport" for the chariot to enter, roughly in the shape of the Jupiter 2, and made of Styrofoam. There were even diorama cutouts, including cutouts of monsters (like the Cyclops!) to fire shells at from the roof of the Chariot.

I just can't imagine a more exciting playset. Unlike most other toys (or even other playsets) based on TV or movie licenses, the Switch 'N Go package allowed you to recreate virtually every key element in the series. Professor Robinson is even given a back pack and a suspension wire to simulate his jetpack flights across the planet, trying to figure out where Will and Penny got themselves lost this week.

Both *Lost* and *Voyage* were to have healthy years in the 1966/67 TV season, but now Irwin Allen was ready to pull what I think in hockey is called a hat rack. He put a third sci-fantasy TV show on the air that season, *The Time Tunnel*.

There weren't any significant toys for *The Time Tunnel*, certainly no figural ones, so it doesn't merit lengthy coverage. It wouldn't merit coverage anyway, since it's a rather repetitive series wherein a couple of one-dimensional characters travel through the fourth dimension and face two-dimensional stock footage. Attempts to liven things up with threats from typical (and recognizable) Irwin Allen aliens came too little, too late. Only lasting that season, this was Irwin Allen's first big TV setback.

Meanwhile, *Voyage* and *Lost* were doing just fine, although *Lost* had definitely, err, lost its way in the story department. Starting toward the end of the previous season, the show had declined into a sitcom format, with Dr. Smith and Will encountering each successive goofball threat of the week, such as Henry Jones as Smith's riverboat gambler-type cousin, and Fritz Feld (Pop!) as an agent of the Intergalactic Department Store.

There were some standout episodes, though. One in particular was called *The Mechanical Men,* in which a group of miniature robots from the planet Industro decide to make the Robinsons' robot their leader. It's easy to see why, they look just like him.

Here's the box for the second-greatest *Lost In Space* collectible of the 1960s, the Robot! The funny-colored outfits worn by the cast in this photo are what they actually wore in the first, black and white season. *Courtesy of The Toyrareum, Ocean City New Jersey.*

Here's how Sears advertised the Robot in '66. Ironically, you can still get it for about 759—only it's dollars today, not cents. *Copyright 1966 Sears, Roebuck, and Co.*

This was accomplished rather economically by using several dozen already-existing toys, specifically, the battery-operated *Lost In Space* Robot from Remco. This fellow, about eleven inches tall, could walk and light up. He was squatter and more rotund than the version on TV, but overall, not a bad likeness. (Ironically, the version from the Switch N Go playset is a bit elongated by comparison.)

The following TV season, 1967-68, *Voyage* ran its fourth year and *Lost* had its third. On *Voyage* it was business as usual, but *Lost* got a major facelift, a jazzy new theme song and new opening credits, new costumes, and major plot changes. Having been stranded on the same planet through most of the first season and a second, identical planet in the second season, the crew of the Jupiter 2 now did much more space-hopping. This was facilitated with the addition of The Space Pod, a mini-ship that they suddenly began using with no explanation and no previous indication of having had. It was so cool, though, that its rather abrupt introduction was quickly for-

Again, we think it's important for you to get a thorough, loving look at this, so here's the back of the box. And again, the box used in these detail shots is *Courtesy of The Toyrareum, Ocean City, New Jersey.*

Here's the bifurcated box top. Never let a toy dealer tell you that it came straight from the factory with one or both flaps already torn off!

given. There was a departure from the show as well. Debbie The Chimp left the show, but went on to star in *Lancelot Link*.

Also, there was greater separation between the drama and comedy elements in each episode, so that the exciting parts seemed genuinely exciting, and the funny parts genuinely funny. Even the really bad episodes were so absolutely asinine that they were enjoyable as solid camp. One time, space hippies (led by Dan Travanti of *Hill Street Blues*) threaten to blow up the Robinsons' current planet. In another episode, Smith turns the Jupiter 2 into a holiday resort for alien fugitives, with help from Fritz Feld ("Fun!" Pop! "Fun!" Pop! "Fun!" Pop!) In short, this was the season of both the best episodes (*The Anti-Matter Man*) and the enjoyably worst (*The Great Vegetable Rebellion*).

Eventually, though, all good things must come to an end. America was tiring of the silly stuff, being generally more interested (in 1968, anyway) with real world concerns like the space race, the race war, and the war in Asia. The summer of 1968 saw the end of Irwin Allen's reign in fantasy TV.

But Irwin's touch was not completely absent from the air in the following 1968-69 season. He adopted the motto of the Round Table by adopting and adapting, though not really improving, his approach to fantasy TV. His new series, *Land Of The Giants*, was set in a pretty normal, accessible world—except that it was being visited by humans from Earth who, relatively speaking, were about six inches high.

Essentially, what I call *Bland Of The Giants* details the adventures of seven colorless characters who spend interminable episodes crawling under doors and through air ducts, yammering at each other on their walkie-talkies, rolling and unrolling coils of string to climb on, and hiding from giant humans and animals who never cast shadows across our heroes. This endless round of captures and escapes would seem a pretty tiresome excuse for entertainment. You would think! And yet, this show lasted the full season and was renewed for the 1969-70 season, the only Irwin Allen show to make it into the 1970s.

The second season highlight was an appearance by Jonathan Harris as The Pied Piper, although he makes little effort to come across as anything other than Dr. Smith. Maybe that's how he was instructed to play it.

Bland succeeded probably because it tapped into the most essential of TV requirements, it just wasn't quite

Chapter Nine

Super Freak-Out Grab Bag!

Like wow, man.

There comes a time in every man's life where he just has to lean back and say, "Where do I put this?" That's why I've selected this final chapter to include everything that didn't quite fit anywhere else. Trying to decide "where to put what" is actually one of the toughest jobs I have in constructing these books.

One thing I tried to do was to limit myself to licensed characters and high-profile figures. Therefore, I've included the Marx 6" Universal Monsters and Super Heroes figurines, but not the Nutty Mads or Campus Cuties. The same goes for Aurora. I have included the models that relate to characters in other chapters, like the handful of spy characters and Irwin Allen characters they did; however, Aurora's extensive super hero and monster lines belong in a book about model kits.

I have also, for the most part, steered clear of battery-operated toys like King Zor, Robot Commando, and Great Garloo. They too belong in another book. On the other hand, I did include Marx's motorized Frankenstein and King Kong because every other Marx licensed character "action" figure is covered in this book.

None of it really makes much sense I realize. But the important thing is that anything that could pass as a play figure in the sense of traditional action figure play IS in here. Okay? Eh.

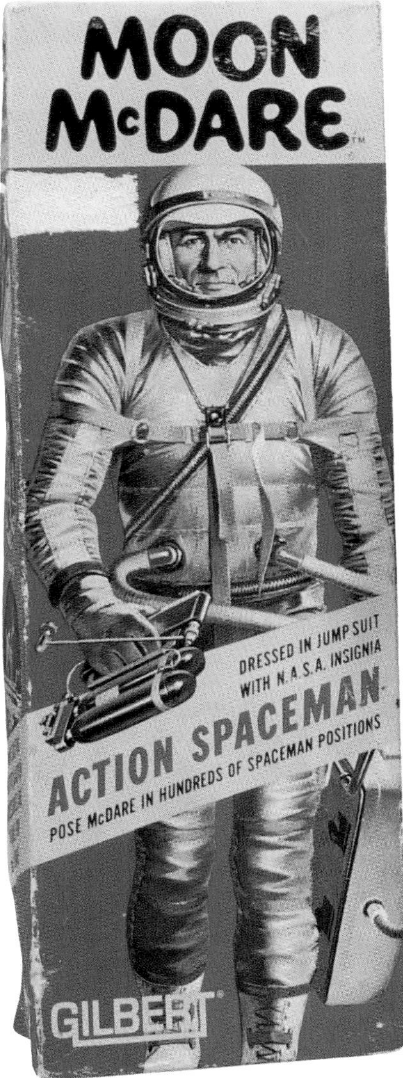

Moon McDare (hyulk) dressed as he came in his box. *Courtesy of Play With This.*

Astronaut Figures

In 1965 Gilbert created Moon McDare as part of its series of a half-dozen action figures, alongside Honey West, "TV's Private Eye-Full" (ouch), James Bond, Oddjob, and the men from U.N.C.L.E. Unlike the other five figures, Moon was Gilbert's own creation and therefore doesn't belong in the Spies chapter. So here he is.

Like the other Gilbert figures, Moon was only jointed at the neck, shoulders, and hips, which made getting him into his space suit extremely difficult.

In a certain sense, Moon outclassed the generally-superior Matt Mason in that the niftiest aspects of the Moon McDare line were the working accessories. His hand-held Geiger counter clicked merrily away, counting all them gammas and betas crawling all over the place. His umbilical cord space line retracted into its housing. He had a working compass. His ray gun buzzed when it was plugged into his backpack. His "space projectile gun" was spring-powered and it fired projectiles not terribly unlike the spears in James Bond's spear gun.

But he had a dumb name, so the heck with him.

Marx's Johnny Apollo, on the other hand, had a cool name. Like other Marx figures, he came fully dressed, in as much as he was, basically, a jointed plastic statue of an astronaut wearing a space suit. He was shorter than most Marx figures, about eight inches tall. But like his taller brothers and sisters, he came with a healthy assortment of accessories.

Left: Moon McDare contemplates wearing white after Labor Day. Don't do it, M.M.!!! *Courtesy of Play With This.*

Below: Rowrf! Space Mutt must release his "satellites" within the confines of his suit. *Courtesy of Play With This.*

Astronaut Figures

Moon McDare	**LMC**	**MIP**
Moon McDare in Jumpsuit	$65	$100
Moon McDare Space Suit	$75	$125
Moon McDare Spear Gun Set	$35	$65
Moon McDare Geiger Counter Set	$30	$55
Moon McDare Space Mutt with Space Suit	$75	$125
Moon McDare Action Communications Set	$60	$95
Moon Explorer Set giant combo deal	$100	$200
Marx NASA Astronaut Action Figure	$50	$95

Moon McDare (hyulk) shared his Sears debut with the *Lost In Space* Robot. Note the Sears exclusive combo of figure, suit, and gun. *Copyright 1966 Sears, Roebuck, and Co.*

A typical Gilbert accessory pack, which conveniently, in this case, is for Moon McDare. *Courtesy of Play With This.*

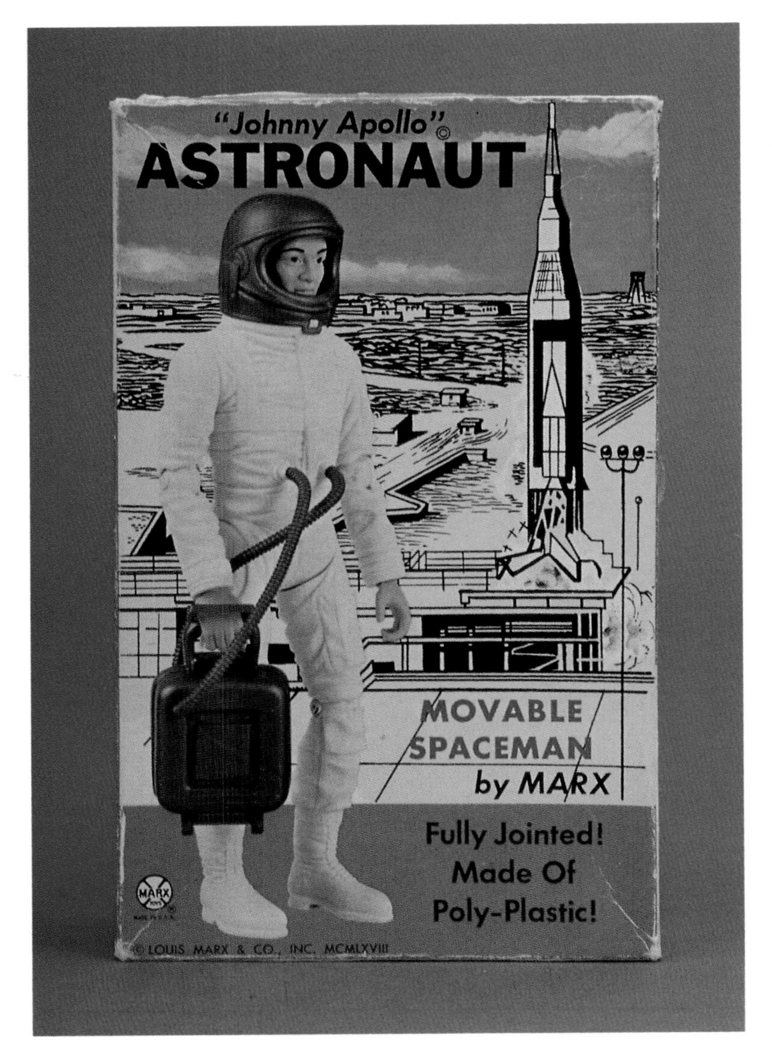

Above: This display-boxed set of air tanks gives you a very good idea of what similar James Bond and *Man From U.N.C.L.E.* sets also looked like. Of course they're much harder to find in unopened packages, since people, err, actually *bought* them. *Courtesy of Play With This.*

Left: The Marx Johnny Apollo NASA Astronaut. Marx put a nice picture of him on his own box, which saved us the trouble of setting him up. Okay, actually, we forgot to get a separate shot of him all set up. God will forgive us. *Courtesy of Play With This.*

Universal Monsters

Aurora's Frankenstein model was produced in 1961 over the objections of many of the company's members. It was a smash hit and in some ways the first major play figure of the 1960s, although it couldn't take much punishment. Because the monster models were such a hit, other companies started to make sturdier versions of these popular stars.

Multiple Products' Pop Top Horrors were unlicensed, but bore a remarkable likeness to Universal's monsters. Each had a removeable head that could be switched around. The heads came attached to the sides of the bodies on sprues, for extra gruesomeness.

Palmer Plastics, on the other hand, obtained the rights and produced some great likenesses. Palmer's line is famous for including characters like Gorgo and the It from It! *The Terror From Beyond Space.*

Universal Monsters	**LMC**	**MIP**
Multiple Products Pop Top Horrors		
Packaged Sets	$40	$75
Unbearably Weird Monsters Multi-Pack	$75	$175
Horror House Target Set	$150	$300
Wolf Man	$20	
Vampire	$20	
Mummy	$20	
Monster (Frankenstein)	$20	
Witch	$15	
Executioner	$15	
Creature of Doom (Reaper)	$15	
Skeleton	$20	

Marx's six-inch monsters, based loosely on the Aurora model kits, were well crafted indeed and looked very much like these. Here's the Mummy! *Courtesy of Play With This.*

Not a real modo, more of a quasi modo, The Hunchback waves to you. *Courtesy of Play With This.*

Palmer Plastics Monster Figurines		
Packaged Sets	$75	$125
Frankenstein	$25	
Dracula	$25	
Wolf Man	$24	
Mummy	$25	
Creature From The Black Lagoon	$40	
King Kong with Faye Wray accessory	$40	
Cyclops	$35	
Gorgo	$40	
It The Terror From Beyond Space	$50	
Marx Monsters		
Battery-Operated Monsters		
Marx Battery-operated grabbing Frankenstein	$300	$500
Marx Battery-operated King Kong	$400	$1000
Marx 6" Plastic Figures		
Frankenstein	$20	
Mummy	$20	
Wolfman	$20	
Hunchback	$15	
Phantom	$20	
Creature From The Black Lagoon	$40	

Here's the Hunchback's hunch! We go for the details. (Except in Johnny Apollo's case.) *Courtesy of Play With This.*

Here's the Phantom. It's all in the wrist. BELLLCH! *Courtesy of Play With This.*

Thanks to Sears, you get a good idea of what the whole Horrible Hamilton deal was. It's really more of a series of playsets than a proper action figure line. But since one of the "lieutenant" bugs was used as the Monster in the *Voyage To The Bottom Of The Sea* playset, it merits this brief mention. *Copyright 1965 Sears, Roebuck, and Co.*

The Sea Devils

While I'd like to say that this was a series based on the monsters from Dr. Who, it was really a tale of futuristic scuba divers and mermen. Made by Mattel, it was in the same vein as Major Matt Mason: 6" bendy men and motorized vehicles.

This is the only figure from the Sea Devils line we were able to snag in time for publication. This was Mattel's underwater equivalent of Major Matt Mason, and it is virtually unknown today. *Courtesy of The Toyrareum, Ocean City, New Jersey.*

Sea Devils Price Guide

	LMC	MIP
Commander Chuck Carter	$25	$95
Rick Riley	$25	$95
Kretor, the fish man	$50	$150
Zark, the motorized whale	$40	$125
Aqualander Vehicle	$25	$85
Search and Rescue Set with Chuck Carter and Aqaulander	$50	$300

Super Heroes

Although they are rare, Ideal's Justice League and Batman playsets are well-documented items. They were toy soldier sized, sold in a variety of sets and packaging styles, and the good guys came painted or unpainted, while the bad guys were solid colors.

But far more rare are the four super hero playsets from Multiple Products, the folks who gave us the Pop Top Horrors. The figures in these sets were not nearly as well sculpted as Ideal's, but all of them, including the villains, were in full color. The highlights of the series are the only 1960s produced figures of The Riddler, Brainiac, Steve Trevor, Angle Man, and The Fisherman. There are also a few made-up characters as well.

The awesome thing about Multiple Product's sets was that there was no mixing; each set came with figures and accessories specific to the main character. The vehicles included in the sets were more or less the products of the imaginations of the fine folks at Multiple Products. See accompanying price guide for complete listings.

This stuff is gold. The Ideal Batman series, all three figures, carded. Notice Joker has his back to us—a typical Joker gag! *Courtesy of The Toyrareum, Ocean City, New Jersey.*

Super Hero Figures

Ideal Justice League and Batman Series

	LMC	MIP
Batman Three-Pack	N/A	$250
JLA Three-Packs	N/A	$250
JLA Four-Packs	N/A	$1500
Superman	$50	
Batman	$50	
Robin	$50	
Wonder Woman	$75	
Aquaman	$75	
Flash	$95	
Joker	$40	
The Key	$40	
Mouse Man	$40	
Brain Storm	$40	
Thunderbolt	$40	
Koltar The 2 Headed Dragon	$40	
Robot	$50	

Unpainted Ideal Batman and Robin loose, without their capes. You'd be lucky to find them at a garage sale even in this shape. *Courtesy of Play With This.*

Ideal's Mouse Man wields a typical weapon for him, a hypo. Of course Mouse Man is supposed to be the size of a mouse, and as such is a rather odd choice for this series. *Courtesy of Play With This.*

I hope our designer Blair puts this shot in real big, because here is most of Ideal's super hero figurine line, all in its natural habitat of the Sears catalog. *Copyright 1966 Sears, Roebuck, and Co.*

Batcave Playset	$1000	$3500
Batcave Alone, with all sections		$750
Batmobile with Seated Figures	$200	
Batplane	$200	
Batman Playset with Batmobile and Villains	$500	$2500
Superman Playset with Car and Figures	$500	$2500
JLA Sanctuary Playset with Heroes and Villains	$3000	$9000
Multiple Products		
Justice League Sets (Canada)		
Superman Set		
(Includes Superman, Superboy, Lois Lane, Brainiac, & Robot Man on tank treads.)	$500	$1000
Batman Set		
(Includes Batman, Joker, Riddler, and Robin on a cycle)	$600	$1200
Aquaman Set		
(Includes Aquaman, Mera, The Fisherman & Mong The Merciless, plus Aqua Sub)	$500	$1000
Wonder Woman Set		
(Includes Wonder Woman, Steve Trevor, Angle Man, Countess Nishki, and Jet)	$500	$1000
Marx Marvel Heroes 6" Statues		
The Incredible Hulk	$20	
The Amazing Spider-Man	$22	
The Invincible Iron Man	$22	
The Mighty Thor	$22	
The Blind Daredevil	$22	
Captain America	$20	
Revell Comic Strip Hero Model Kits		
The Phantom and the Witch Doctor	$100	$300
Flash Gordon and the Martian	$100	$300

Marx's six-inch Captain America, painted. *Courtesy of The Toyrareum, Ocean City, New Jersey.*

Marx' Mighty Thor, painted. *Courtesy of The Toyrareum, Ocean City, New Jersey.*

Marx's six-inch Captain America—all together now—UNpainted! *Courtesy of Play With This.*

Smaller unmarked Thor. *Courtesy of The Toyrareum, Ocean City, New Jersey.*

Ablaze with power, he's Marx's six-inch Iron Man! *Courtesy of Play With This.*

Let's level with Daredevil. *Courtesy of Play With This.*

Zeroids

Zeroids were robots created by Ideal to be in scale with Mattel's popular Major Matt Mason series. Although robot toys are not generally considered action figures, these were designed to interact with each other and use their equipment, so as far as I'm concerned they're battery-powered robot action figures.

Now, here was a cool action figure line. Not only were they designed in the tradition of the lovable, rotund robotic rowdy on *Lost In Space*, not only did they fill the "robot gap" left by the more organic Major Matt and Outer Space Men, but also they were motorized and actually did things. Their motors allowed them to move around, operate their vehicles, and so on. Each of the Zeroids even came in a package that doubled as an accessory. Now how cool is THAT?

Zerak, the Blue Destroyer, came in a display case with a fold-down ramp. Zobor, the Bronze Transporter, had a case that served as a space cart vehicle. Zinthar, The Silver Explorer, had a case that doubled as a space sled. There was also a Solar Cycle packaged with a Zeroid figure. Every Zeroid figure was offered individually with the Solar Cycle.

Not long after came the green Zeroid Commander, Zogg. He was available with the motorized Zem 21 spaceship, which mechanically released a ramp that allowed Zogg to come charging out of the ship to attack Major Matt Mason. There was also the Zeroid Commander Action Set, with its Solar Cycle and a Sensor Station with revolving antenna and sonic alarm to warn Zogg of approaching Outer Space Men.

If you didn't have enough Zoggs, you could get him again in the boxed deluxe set which included also the original three Zeroids and an evil Alien who flew apart when attacked ... and was only available in this playset.

An altered, generic Zeroid and space ship were rereleased when Ideal issued its Star Team series in the 1970s, a strange hybrid that included a Darth Vader clone with Captain Action's boots and body. For more info on the Star Team see *G.I.Joe™ And Other Backyard Heroes*.

The Bronze Transporter Zeroid!
Courtesy of Play With This.

Zeroids (IDEAL 1968-1970)	**LMC**	**MIP**
Blue	$75	$95
Bronze	$75	$95
Silver	$75	$96
Commander Zogg Playset	$300	$600
Commander Zogg (Metallic Green)	$75	
Alien: Green, Fly-Apart	$200	
Five-Figure Set	$350	$800

And that about wraps it up. If you haven't already, run out and get the awesome 1970s book *G.I.Joe™ And Other Backyard Heroes 1970-1979*, and the third volume in this series, *Action Figures Of The 1980s*. I wrote them both and they're as wacky as this one.

Collect 'Em All!!!

Zobor here is missing the antenna from atop his head, proving that even packaged Zeroids can have easily-lost parts. *Courtesy of Play With This.*

Commander Zogg, the ultimate Zeroid. *Courtesy of The Toyrareum, Ocean City, New Jersey.*

Zogg from another angle. Woo Hoo! *Courtesy of The Toyrareum, Ocean City, New Jersey.*

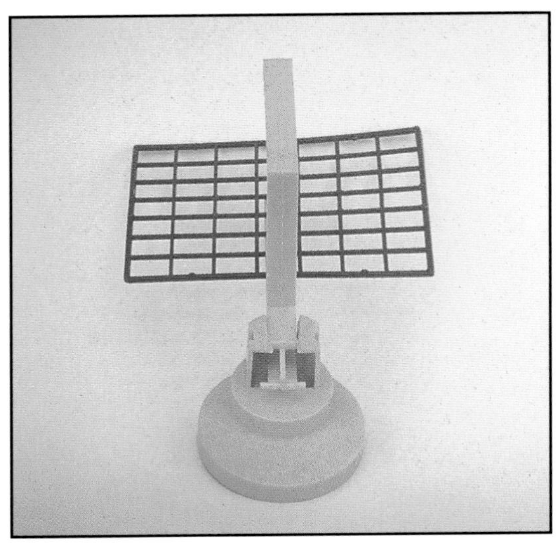

The Radar Dish from Zogg's action set. Somebody assembled it wrong, but the pieces are all there. All Action Set pieces are *Courtesy of The Toyrareum, Ocean City, New Jersey.*

The yellow ramp thing from what is officially known as the Zeroid Commander Action Set.

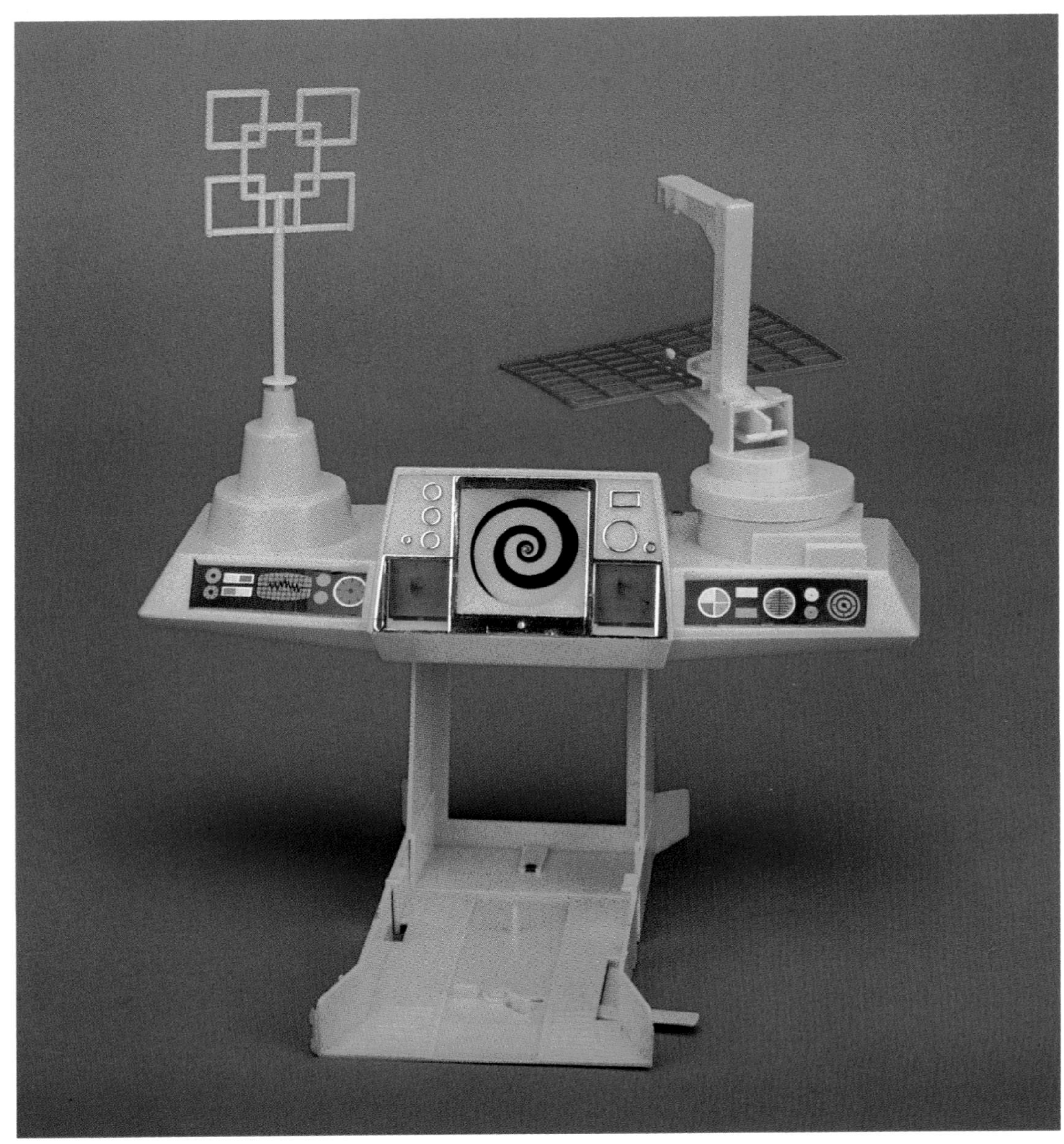

The ultra cool Zeroid Command Console.

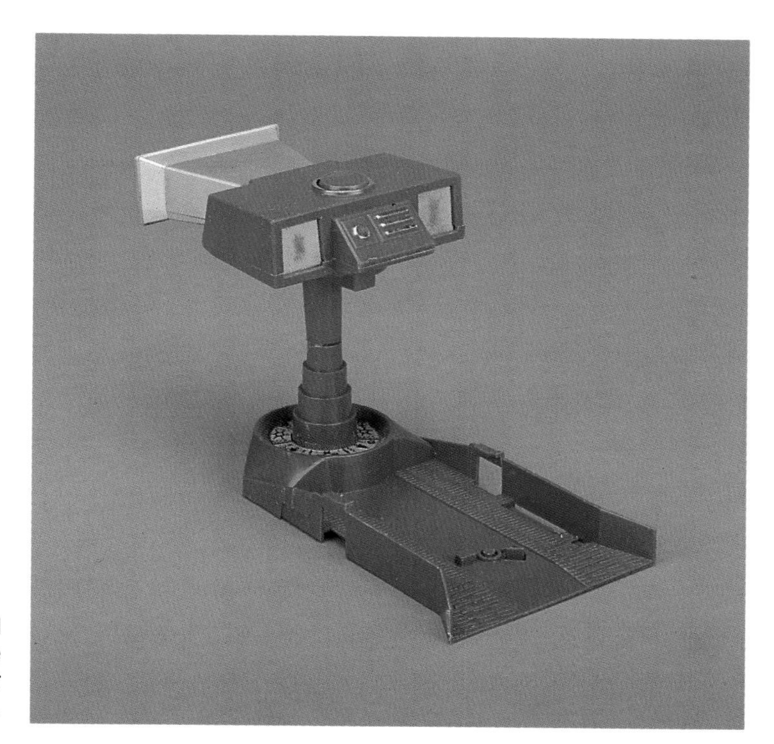

The blue sled from the Commander Playset.

The Big Wheel from the Commander Playset, which Zogg gleefully demonstrates for you.

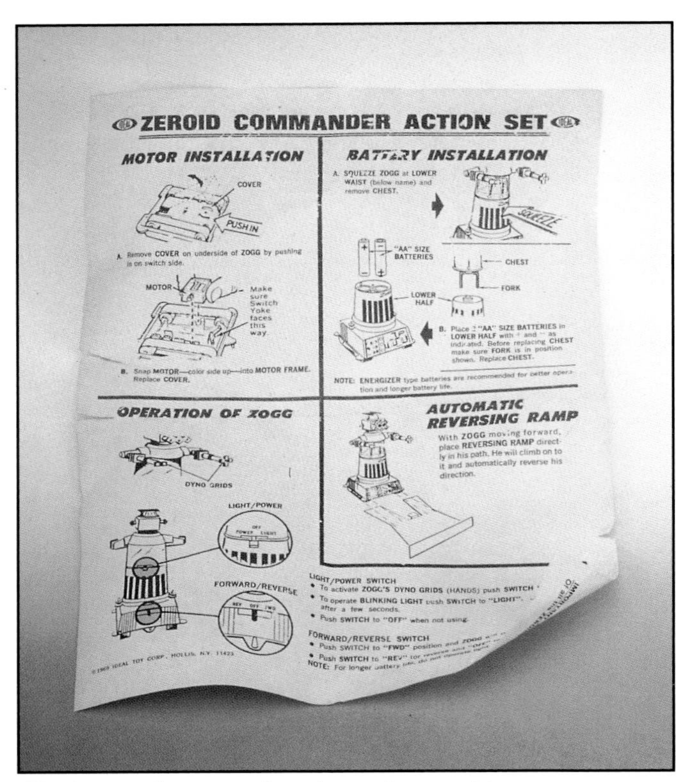

ZEROID COMMANDER ACTION SET

MOTOR INSTALLATION

A. Remove COVER on underside of ZOGG by pushing
is on switch side.

B. Snap MOTOR—color side up—into MOTOR FRAME.
Replace COVER.

BATTERY INSTALLATION

A. SQUEEZE ZOGG at LOWER WAIST (below name) and remove CHEST.

B. Place 2 "AA" SIZE BATTERIES in LOWER HALF with + and − as indicated. Before replacing CHEST make sure FORK is in position shown. Replace CHEST.

NOTE: ENERGIZER type batteries are recommended for better operation and longer battery life.

OPERATION OF ZOGG

AUTOMATIC REVERSING RAMP

With ZOGG moving forward, place REVERSING RAMP directly in his path. He will climb on to it and automatically reverse his direction.

LIGHT/POWER SWITCH

- To activate ZOGG'S DYNO GRIDS (HANDS) push SWITCH
- To operate BLINKING LIGHT push SWITCH to "LIGHT". after a few seconds.
- Push SWITCH to "OFF" when not using.

FORWARD/REVERSE SWITCH

- Push SWITCH to "FWD" position and ZOGG
- Push SWITCH to "REV" for

NOTE: For longer battery life,

The instructions—again, where else would you see—okay, okay. *Courtesy of The Toyrareum, Ocean City, New Jersey.*

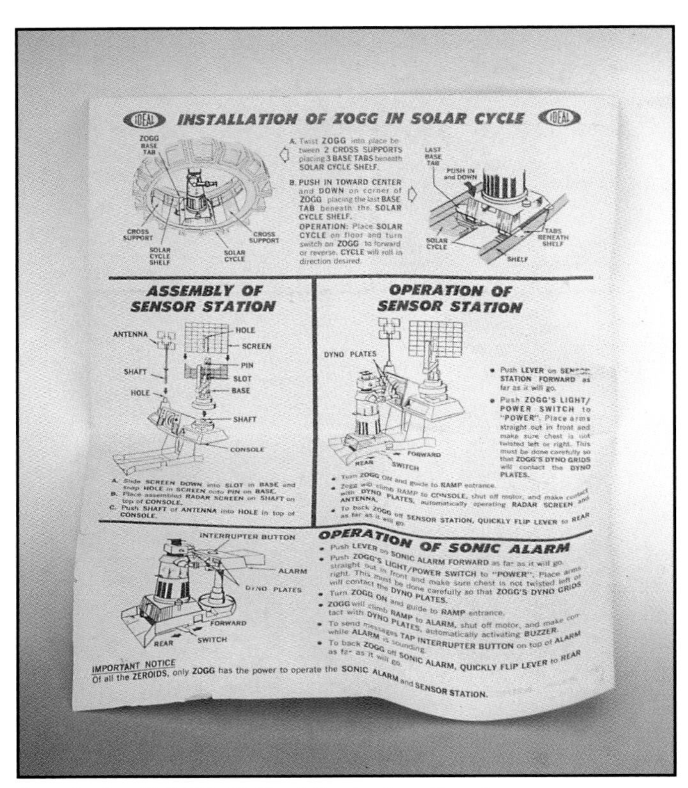

INSTALLATION OF ZOGG IN SOLAR CYCLE

A. Twist ZOGG into place between 2 CROSS SUPPORTS placing 3 BASE TABS beneath SOLAR CYCLE SHELF.

B. PUSH IN TOWARD CENTER and DOWN on corner of ZOGG placing the last BASE TAB beneath the SOLAR CYCLE SHELF.

OPERATION: Place SOLAR CYCLE on floor and turn switch on ZOGG to forward or reverse. CYCLE will roll in direction desired.

ASSEMBLY OF SENSOR STATION

A. Slide SCREEN DOWN into SLOT in BASE and snap HOLE in SCREEN onto PIN on BASE.
B. Place assembled RADAR SCREEN on SHAFT on top of CONSOLE.
C. Push SHAFT of ANTENNA into HOLE in top of CONSOLE.

OPERATION OF SENSOR STATION

- Push LEVER on SENSOR STATION FORWARD as far as it will go.
- Push ZOGG'S LIGHT/POWER SWITCH to "POWER". Place arms straight out in front and make sure chest is not twisted left or right. This must be done carefully so that ZOGG'S DYNO GRIDS will contact the DYNO PLATES.
- Turn ZOGG ON and guide to RAMP entrance.
- Zogg will climb RAMP to CONSOLE, shut off motor, and make contact with DYNO PLATES, automatically operating RADAR SCREEN and ANTENNA.
- To back ZOGG off SENSOR STATION, QUICKLY FLIP LEVER to REAR as far as it will go.

OPERATION OF SONIC ALARM

- Push LEVER on SONIC ALARM FORWARD as far as it will go.
- Push ZOGG'S LIGHT/POWER SWITCH to "POWER". Place arms straight out in front and make sure chest is not twisted left or right. This must be done carefully so that ZOGG'S DYNO GRIDS will contact the DYNO PLATES.
- Turn ZOGG ON and guide to RAMP entrance.
- ZOGG will climb RAMP to ALARM, shut off motor, and make contact with DYNO PLATES, automatically activating BUZZER.
- To send messages TAP INTERRUPTER BUTTON on top of ALARM while ALARM is sounding.
- To back ZOGG off SONIC ALARM, QUICKLY FLIP LEVER to REAR as far as it will go.

IMPORTANT NOTICE
Of all the ZEROIDS, only ZOGG has the power to operate the SONIC ALARM and SENSOR STATION.

Here's the OTHER SIDE of the instructions!

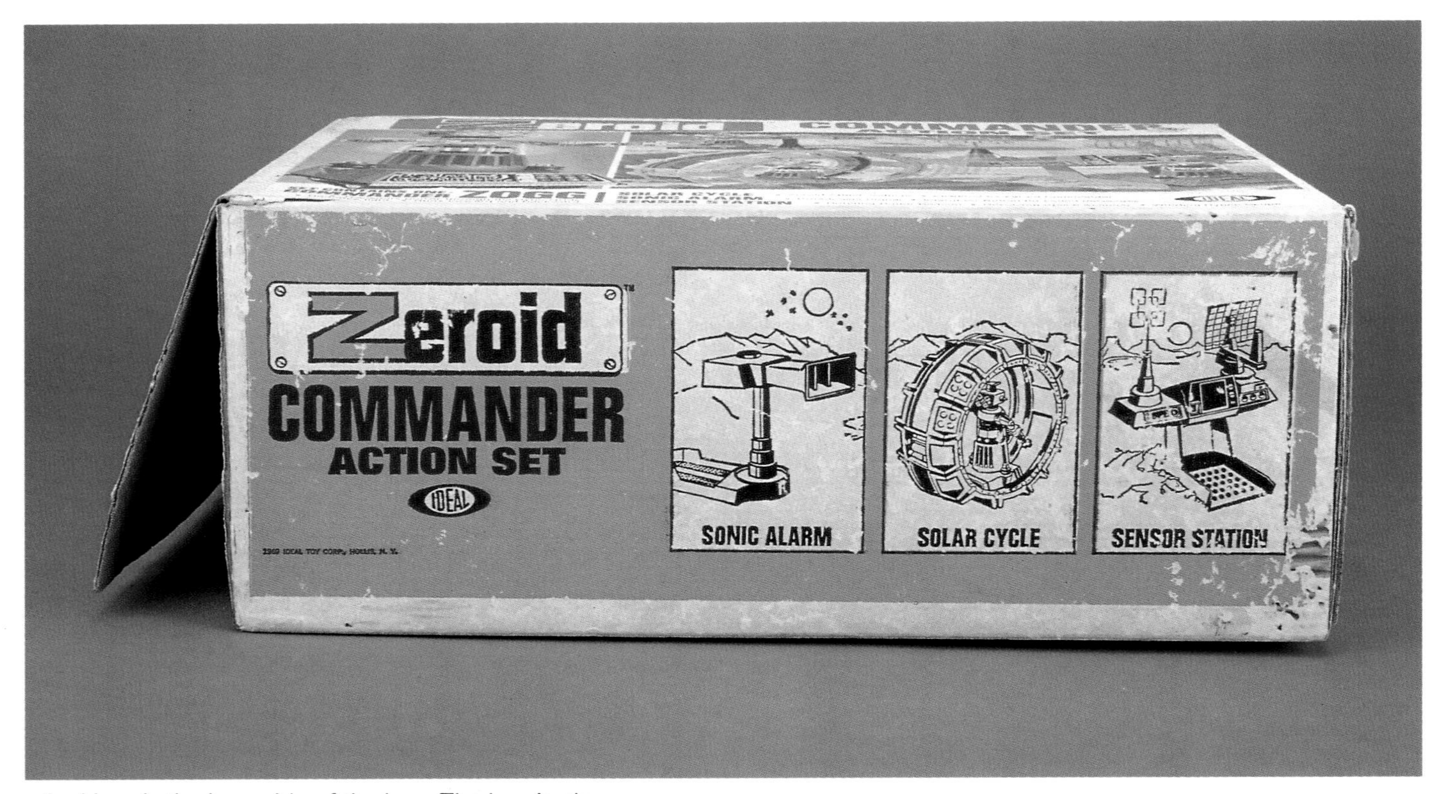

And here's the long side of the box. The box is also *Courtesy Of The Toyrareum, Ocean City, New Jersey.*

And here's the short side of the box!

And here it is, the FRONT OF THE BOX!
Look at the colors! Woo Hoo! And now...

It's Dr. Evil in a dress, for our big finale! Thank you Fred Mahn for the loan of your sartorially somnambulistic emissary of evil! And thank you, 1960s, for all your cool toys! Cowabunga, Begorah!